THE WALDORF APPROACH TO ARITHMETIC

Hermann von Baravalle

edited by David Booth

A Science and Mathematics Association
for Research and Teaching Book

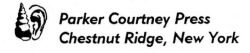 Parker Courtney Press
Chestnut Ridge, New York

Contents

Biographical Notes

Hermann von Baravalle (1898-1973) was one of four children in an aristocratic family that lived across the street from the castle of the Kaiser of the Austro-Hungarian Empire. He lived in a protected environment as a child but was swept into the dramatic changes brought by the twentieth century when he volunteered for duty in World War I, at the age of seventeen.

One of his instructors in the officer's academy was W. J. Stein. He learned of anthroposophy, the philosophy developed by Rudolf Steiner, from Stein. His first furlough from the front was used to hear Rudolf Steiner speak in Berlin.

Eventually he was chosen by Rudolf Steiner to teach in the first Waldorf School which opened in Stuttgart, Germany, in 1919; Stein, his friend from the Austrian army, was a colleague there. Before taking up the post, however, he completed his Ph.D. at the University of Vienna, using the opportunity to develop some ideas concerning education in physics and mathematics. He took up teaching math and physics in the second year of the Waldorf School. For sixteen years he taught there and at its associated Teacher Training Seminar. As a teacher he insisted on punctuality, was very demanding about the quality of the handwriting that went into the pupil's notebooks, and took pains to uphold high artistic standards in his lessons. His enthusiasm for mathematics was contagious. The pupils genuinely enjoyed mathematics lessons.

Dr. von Baravalle's career as a lecturer began simultaneously with his teaching post. He gave three lectures at the opening of the first Goetheanum in Dornach, Switzerland, when he was 22 years old. Rudolf Steiner had urged him to perfect his command of English. He married Erica Brown, an English woman, who was a successful English teacher to the children in the younger grades of the Waldorf school. They had one son.

In 1933, in the company of Ehrenfried Pfeiffer and Guenther Wachsmuth, Dr. von Baravalle undertook a trip to America. Lisl Franceschelli-Berlin, one of his pupils, remembers the stories that he told to his class about his

American journey. Bridges, railways, and large buildings that he had seen in America reappeared as mathematical problems for his students.

When the German government brought all education under its domination in the years prior to the Second World War, the Waldorf schools were closed: Dr. von Baravalle was invited to England and then to America. He arrived in 1937 and became an American citizen in 1944.

He taught for a time at the Edgewood School in Greenwich, Connecticut, before its metamorphosis into High Mowing School of New Hampshire in 1942; he was the educational advisor to the Kimberton Farms School (now the Kimberton Waldorf School) when it opened in 1941 near Philadelphia.

In 1943, Dr. von Baravalle was appointed Professor of Mathematics and Chairman of the Mathematics Department of Adelphi College in Garden City, Long Island. During this period he wrote articles for mathematics teachers and was hopeful that the teaching profession in American schools would make good use of the experience acquired before the war in the European Waldorf schools.

In 1946 a Waldorf Teacher Training program was begun at Adelphi College and he became the Educational Director of the Waldorf Demonstration School of Adelphi, which later became the Garden City Waldorf School.

In the early 1950's he advised the Waldorf Schools that reappeared across West Germany soon after that country began its post-war recovery and was closely involved in the foundation of several new American schools as well. Gisela O'Neil, whose lecture notes appear in this volume, was among the first teachers at Highland Hall School when it began in Los Angeles, with the aid of Dr. von Baravalle in 1955. Meanwhile, he lectured before anthroposophical groups and mathematics educators. Many people have a vivid memory of these lectures. "Magician" is a word sometimes used to describe the impression he made on audiences during his mathematical talks.

In 1969 Dr. von Baravalle suffered a stroke. He entered a nursing home in Germany and died there in 1973.

The Four Sections of this Book

A volume on arithmetic will, of course, lack the exquisite geometrical drawings for which Hermann von Baravalle was famous. It is his arithmetical lectures, however, that are most urgently needed in the classroom.

Psychological Points of View on the Teaching of Arithmetic is the opening article. It seems to have been von Baravalle's first publication in English and serves well as an introduction to his lectures on number theory. It is printed here with the permission of the *Mathematics Teacher* where it first appeared. A few minor changes have been made so that the text is more natural in English.

Lectures on Number Theory is a rich source of ideas for teachers. I know from experience that many sections can be adapted for successful use in grade school and high school classes. There is, however, no explicit script for teachers to follow. These lectures were used to train prospective class teachers for Waldorf schools; the teachers were to teach out of their own understanding. The course, however, is not just theoretical: It often shows Dr. von Baravalle's own classroom experience.

The Number π is a finished article. The subject matter is of philosophical as well as mathematical importance. It describes a grand epic of human reason that has a simple theme at its core: The number π is defined geometrically in a very natural way, but how can it be captured in arithmetical terms?

The last section of this volume, *Notes on Waldorf Education Part I*, is taken from lecture notes. We have not included the entire course of lectures here. His remarks on the alphabet and his geometrical observations have been left out and reserved for a future volume on school geometry for which they are better suited.

A thorough bibliography of von Baravalle's writings appears at the back of the book. Henry Saphir has contributed to this bibliography and other editorial tasks as well.

David Booth   Green Meadow Waldorf School

Psychological Points of View
on the Teaching of Arithmetic

The psychological point of view focuses primarily on the effect that the learning of a subject has on the mind itself. It seeks to discover the creative impulses that can be conveyed, through the lessons of arithmetic during the various grades.

Already with children at an early age, even before they enter school, phenomena due to creative contact with numbers can be observed. When the child begins to count he becomes aware of the fact that his use of the words one, two, three, four, etc. is different from his usual speech. They have their regular sequence. The word *one* always comes first in counting. The word *two* has to wait until *one* has passed and *three* must wait even longer, etc. Each one of these little words is preceded by another one and in its turn it is the key word for the following one. If a child is given the opportunity to count out loud repeatedly, the impression of this sequence of words is strengthened and a joyful response is awakened. The fact that the child hears himself speak connectedly in counting calls forth the satisfaction of moving within an ordered sequence. This satisfaction does not derive from indulging in subjective arbitrariness but from partaking in lawfulness.

From simple counting arithmetic lessons can proceed in the first grade to rhythmic counting, accentuating some numbers in contrast to others. This can be done by speaking them louder or by pronouncing them more distinctly and slowly in comparison to others passed over more quickly. As an example, accent every third number when counting and the sequence obtained is

1 2 **3** 4 5 **6** 7 8 **9** 10 11 **12**

By accenting every fourth number the following sequence results:

1 2 3 **4** 5 6 7 **8** 9 10 11 **12** 13 14 15 **16**

Multiplication tables are obtained by making a choice among the complete sequence of numbers, that is dropping some and retaining others in rhythmic order. To the child's first impression of the set order in numbers there now comes an additional one, that of making regulated choices from their original sequence. In connection with these exercises the pupils can walk or clap their hands in rhythm with the counting, or perform any similar motion. Introducing the element of rhythm into arithmetic brings it close to the experience the child has had in music.

In the historical development of mathematics during the Greek period mathematical rhythmic procedures were already applied. Eratosthenes, for instance, when determining the prime numbers, went repeatedly over the regular sequence of numbers and eliminated first every second number that followed two thus:

$$1 \ 2 \ 3 \ \cancel{4} \ 5 \ \cancel{6} \ 7 \ \cancel{8} \ 9 \ \cancel{10} \ 11 \ \cancel{12} \ 13 \ \cancel{14} \ 15 \cdots$$

Then he eliminated also every third number that followed 3, thus:

$$1 \ 2 \ 3 \ \cancel{4} \ 5 \ \cancel{6} \ 7 \ \cancel{8} \ \cancel{9} \ \cancel{10} \ 11 \ \cancel{12} \ 13 \ \cancel{14} \ \cancel{15} \cdots$$

When taking up four, the whole row is found already eliminated. By crossing out every fifth number that follows 5 some numbers, like 10, 15, and 20, are found already eliminated. The first number remaining to be crossed out is 25. The numbers that remain after the whole procedure has been completed are the prime numbers.

Another view on arithmetic from the psychological angle is derived from the very nature of numbers. Numbers lead the activity of the mind beyond immediate perception. We do not perceive numbers as sense phenomena, as we do colors, sounds, etc.; numbers arise in our mind in addition to sense observation. This fact becomes apparent when we recognize the motive behind an impetus to count. The leaves on the branch of a tree, for instance, would not excite in us a desire to count them – the result obtained would not be worth the effort. The case is different with the petals of a flower, for their number is often characteristic of the blossom and determines its shape. What makes it reasonable to count in one case and not in the other is the existence or lack of a relationship between the elements that we count. This is recognized in the fact that apples and pears can be added only after a relationship between apples and pears has been established under the common concept *fruit*.

The process of counting can be studied by emptying a basket of apples on a table. Involuntarily with the hand, or only in our thoughts, we arrange them in groups and count them as 3 + 4 + 2 or 3 + 3 + 3. Nine objects, lying about on the table, can then be counted at the first glance. In the following the 9 points are arranged first without order, then as a 3 × 3 square:

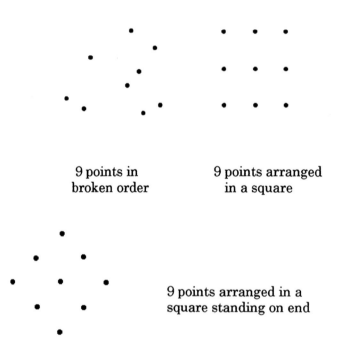

9 points in   9 points arranged
broken order   in a square

9 points arranged in a
square standing on end

The successive horizontal lines contain, in the square standing on end, 1 + 2 + 3 + 2 + 1 points. Similarly, four times four points lead to the following result.

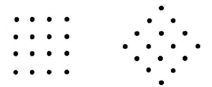

16 points arranged   16 points arranged in
in a square   a square standing on end

$$16 = 1 + 2 + 3 + 4 + 3 + 2 + 1.$$

And 5 times 5 points gives

$$25 = 1 + 2 + 3 + 4 + 5 + 4 + 3 + 2 + 1$$

Summing up the successive square numbers one obtains

$$1 \times 1 = 1$$
$$2 \times 2 = 1 + 2 + 1$$
$$3 \times 3 = 1 + 2 + 3 + 2 + 1$$
$$4 \times 4 = 1 + 2 + 3 + 4 + 3 + 2 + 1$$
$$5 \times 5 = 1 + 2 + 3 + 4 + 5 + 4 + 3 + 2 + 1$$

In each row the numbers increase up to the middle and then decrease symmetrically. The square of every number represents the sum of the arithmetical progression of numbers going from one up to the original number and back again to one. Every subsequent line has 2 numbers more than the preceding one; from line to line a higher middle number is added, and the number preceding it is repeated. The new numbers thus added are

$$1 + 0 = \quad 1$$
$$2 + 1 = \quad 3$$
$$3 + 2 = \quad 5$$
$$4 + 3 = \quad 7$$
$$5 + 4 = \quad 9$$
$$6 + 5 = \quad 11$$
$$7 + 6 = \quad 13$$
$$8 + 7 = \quad 15$$
$$\cdot \; \cdot \; \cdot \; \cdot \quad \cdot \; \cdot$$

The odd numbers manifest themselves as the differences between the square numbers:

$$1 - 0 = \quad 1$$
$$4 - 1 = \quad 3$$
$$9 - 4 = \quad 5$$
$$16 - 9 = \quad 7$$
$$25 - 16 = \quad 9$$
$$36 - 25 = \quad 11$$
$$49 - 36 = \quad 13$$
$$64 - 49 = \quad 15$$
$$\cdot \; \cdot \; \cdot \; \cdot \quad \cdot \; \cdot$$

Many number relationships can be demonstrated to arouse the pupils' interest in mathematical facts. They appeal no less to the mind than experiences of color and form and hold the additional stimulus that they can be revealed without any other tool than the mind's own activity.

In working with a class special occasions may be seized to display numerical facts. On a certain day, one of the children may have his eighth birthday. His little brother at home may then be 2 years old. The child therefore lived four times as long as the little brother. One may further call attention to the fact that in the next year the little child will have his ninth birthday and the little brother his third. The age ratio between them has changed; the child has now three times the little brother's age. A year later the child will be ten and the little brother four, then eleven and the little brother five, the next year twelve and the brother six. At this point the child's life will be twice that of the brother. This diminished ratio will make the child think, and it can be pointed out to him that this ratio also becomes manifest in life. At the time when the child is eight and has four times the age of the little brother, the age difference between the two children is apparent to everyone on sight. But once they have reached 42 and 48 it might possibly be doubtful to recognize which was the older of the two.

An important factor from a psychological point of view is the continuity in the thoughts pursued during the lesson. If the problems introduced in a lesson deal, for instance, first with stretches traversed by trains, then with pipe lines that fill a container and afterwards with the sales price of fruits and vegetables, etc., the continuity of thought is disrupted each time. By continuing a line of thought greater interest can be aroused in both verbal and numerical problems. For instance, after taking up $12 = 5 + 7$, one can proceed to elaborate further on the problem by giving other compositions of this number, thus:

$$
\begin{aligned}
12 &= 1 + 11 & 12 &= 7 + 5 \\
12 &= 2 + 10 & 12 &= 8 + 4 \\
12 &= 3 + 9 & 12 &= 9 + 3 \\
12 &= 4 + 8 & 12 &= 10 + 2 \\
12 &= 5 + 7 & 12 &= 11 + 1 \\
12 &= 6 + 6
\end{aligned}
$$

The first number to the right of the equal sign increases step by step from one to eleven, while the one in the next column diminishes from eleven to one. The amount of possible combinations is eleven, one less than the original number 12. Continuing this problem one can follow up with the factoring of the same number:

$$
\begin{aligned}
12 &= 1 \times 12 \\
12 &= 2 \times 6 \\
12 &= 3 \times 4 \\
12 &= 4 \times 3 \\
12 &= 6 \times 2 \\
12 &= 12 \times 1
\end{aligned}
$$

The amount of possibilities in the case of factoring varies with the number chosen:

$13 =\ \ 1 \times 13$	$14 =\ \ 1 \times 14$	$15 =\ \ 1 \times 15$
$13 = 13\ \times\ 1$	$14 =\ \ 2\ \times\ 7$	$15 =\ \ 3\ \times\ 5$
	$14 =\ \ 7\ \times\ 2$	$15 =\ \ 5\ \times\ 3$
	$14 = 14\ \times\ 1$	$15 = 15\ \times\ 1$

With thirteen there are only two possibilities, while with twelve there are six. The numbers 14 and 15 offer four possibilities each. The children will seek the numbers with the most possibilities. Naturally, as a rule, the number of possibilities will increase when higher numbers are reached. In order to judge which number has a relatively great amount of possibilities we can investigate how far one has to proceed from it before encountering one with more possibilities. One of the numbers with many factoring possibilities is 12. Any number below it offers no more than 4 possibilities, while 12 itself offers 6. Continuing from 12 the amount of possibilities is not exceeded until 24 is reached. In the range from 1 to 100 the number 60 reaches the maximum of 12 factoring possibilities, which is not exceeded by any one of the following numbers including 100 itself. This fact explains the predominating role of the numbers 12 and 60 in our units of measurement. The clock shows the number 12; the year has 12 months; the foot 12 inches, etc. An hour has 60 minutes, a minute 60 seconds, the 60th part of $\frac{1}{6}$ of the circumference of a circle equals 1°, etc.

Much that belongs to a later period of study can be prepared for at an earlier one. For instance, when taking up even and odd numbers their addition can be expressed in the following rule:

Even number + even number = even number
Even number + odd number = odd number
Odd number + even number = odd number
Odd number + odd number = even number

In later years when the subject of negative numbers is introduced a similar rule occurs:

Positive number × positive number = positive number
Positive number × negative number = negative number
Negative number × positive number = negative number
Negative number × negative number = positive number

This apparent analogy is not accidental but is mathematically founded (even powers of a negative are positive, odd powers negative).

When practicing long multiplication and division, one may prepare for the subject of roots. Instead of taking an arbitrarily chosen number for practice, one may multiply the number 1.4142136 with itself. The result is 2 followed by 6 zeros: The number 1.4142136 is the square root of 2 (up to 7 decimals), the last decimal being rounded upward so that the result of the multiplication is above 2. Then multiplying 1.4142135 with itself the result is 1 followed by a series of nines; that is a result below 2. A similar exercise with 1.73206 and 1.73205 gives results above and below 3; by multiplying 2.23607 and 2.23606, each with itself, one result is higher and one lower than 5, etc. multiplying

$$2.15445 \times 2.15445 \times 2.15445$$

we get 10 followed by three zeroes, etc. Then in later years when the study of roots is taken up ($\sqrt{2} = 1.4142135\cdots$; $\sqrt{3} = 1.73205\cdots$; $\sqrt{5} = 2.23606\cdots$; $\sqrt[3]{10} = 2.15443\cdots$), it will be an easy matter to proceed on this foundation. But even more important than the advantages thus acquired for the progress of studies is the almost unconscious psychological effect. The pupil has the experience that his study course since his early years been built up with a deeper background than he could appreciate, and this experience creates a feeling of confidence.

The way of intimating the more advanced portion of the subject in the early lessons can be extended even beyond the respective school. For instance, when practicing the four rules of arithmetic in the early years the following scheme may easily be set up:

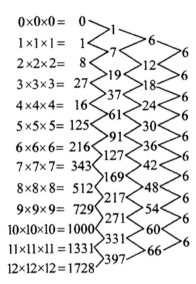

First find the number which results from multiplying a number with itself three times, the cube numbers. In the next column write down the differences between the successive cube numbers, their 1st differences. In the next column are the differences of the differences: 2nd differences. Finally, the third differences have the constant value of 6. This fact is fundamental in the differential calculus. Perhaps ten years later the student may come across a similar situation when he will have advanced in his studies.

Much can be done to arouse the interest in mathematical problems by reciprocal comparison of different kinds of numbers and methods of calculation. For instance, when introducing fractions, compare their nature with that of regular numbers, not theoretically but in the form of problems. The differences between successive fractions are not equal as they are with regular numbers, but diminish continually. The step from 1 to 1/2 has a difference of 1/2; that from 1/2 to 1/3 a difference of 1/6, etc. In the series of numbers

$$1 \quad 2 \quad 3 \quad 4 \quad 5 \quad 6 \quad 7$$

the arithmetical mean between 2 and 6 is: $(2 + 6)/2 = 4$. In the series of fractions

$$1 \quad \frac{1}{2} \quad \frac{1}{3} \quad \frac{1}{4} \quad \frac{1}{5} \quad \frac{1}{6} \quad \frac{1}{7}$$

the mean between 1/2 and 1/6 is:

$$\frac{\frac{1}{2} + \frac{1}{6}}{2} = \frac{1}{3}$$

The mean 4 of 2 and 6 stands in the middle of 2 and 6, whereas in the series of fractions the mean of 1/2 and 1/6 is not 1/4 but 1/3, which no longer stands in the middle, but directly follows the first number 1/2. The sum of the differences between 1/3 and 1/4, 1/4 and 1/5, 1/5 and 1/6 is equal to that between the successive fractions 1/2 and 1/3.

This shows why the simple rules for addition and subtraction, applied in the case of integers, are no longer valid with fractions, and that another procedure, including the finding of the common denominator, has to be used. The fractions diminish in a definite way. In order to grasp this more readily have the children hold their hands at a fixed distance from each other. This will express the unit. Then have them move their hands closer together step by step, so that the intervals gradually diminish to 1/2, 1/3, 1/4, 1/5, 1/6···. The differences here grow less and less. Now it is but a step to graph the successive fractions in the form of heights over a common base line resulting in a branch of an hyperbola. This graph appears again as the diagram of Boyle's law in physics; in other words, the opportunity is here given to make allusion to geometry and physics which will be further developed in the later school years.

What is of special importance from the psychological point of view for all school grades is the quality of thinking activity which is provoked by the methods employed. The result of arithmetic classes may be either a mechanical and primitive thinking or an intensive and creative one. There are students who when given a mathematical problem will at once ask for a fixed rule or formula with which to solve it. Their work is not based on insight. There is no continuity in their consciousness just where the most real thought connection should set in. They usually have primitive, mathematical concepts, conceiving any length as the sum of so and so many inches, any weight as the sum of so and so many grams, etc. But neither a length nor a weight is composed of parts; each forms a whole that is to be compared with others. A more intensive thinking will be more apt to aim at comparisons, at proportional numbers and less at dissecting and combining. If a measure is three times that of another, this fact remains whether the measuring be done in meters or inches. Nearly all the numbers used in the sciences are fundamentally proportional numbers. If a stone is said to have a specific gravity of 3, that means that its weight is three times that of an equal volume of water; it is not a combination of three units.

Psychological effects produced by various kinds of mental activities are naturally not limited to their respective realms. They continue to act in the conscious as well as in the subconscious spheres of our mind and their influence finally affects even social life. No healthy relationship and cooperation between human beings can be founded on mechanical and one-sided thinking. Such thinking isolates one within oneself and builds no real link toward the understanding of the other person's point of view. Thus such thinking is often the cause of dissatisfaction and lack of self-confidence. It leads to an attitude of self-defense toward life without courage or initiative, or to a uselessly critical and cynical attitude.

The psychological effect of creative studies in arithmetic is of at least equal importance as is the knowledge the subject imparts. Whoever looks only for the "practical" side in arithmetic readily overlooks the truly practical effects within the psychological realm. Particularly in our days this is becoming increasingly essential. The dangers threatening culture and civilization today are not primarily due to any lack of detailed practical knowledge, but rather to psychological factors which, in the final analysis, are the outgrowth of distorted thinking. The teaching of arithmetic in its education of sound and straight and creative thinking holds a key position in our work for a future society throughout the world.

The Theory of Numbers

Notes on the Lectures of Herman von Baravalle
Taken by Gisela Thomas O'Neil

Introduction

Numbers are not just counters standing in ranks, row upon row, to march forward and serve our enumerative purposes. They have qualities that go beyond that of their position in the familiar ordering. In order to become more intimately acquainted with the integers, we need to familiarize ourselves, first of all, with their multiplicative relationships.

These notes come from a course that was designed for the training of teachers. The topics are well-chosen for teacher education. They are of historical interest; they support each other, are based in computational experience, and they leave us feeling that arithmetic is alive. The lectures begin with an approach to higher arithmetic that is simultaneously rhythmic and rational.

The formulas, calculations, and order of presentation in these notes are those of the lecturer. The verbal explanations, however, have been added. Without the added text, the abbreviated notes would have been a source of numerous puzzles which every reader would have to solve anew. There has been no attempt to reproduce von Baravalle's style. Our aim was to make the concepts clear while avoiding redundant, explanatory text.

The lectures seem to have been given at Adelphi College in about 1950 – the notes themselves were not dated. In several places von Baravalle's lectures resemble the treatment used in the book *Number Theory and Its History* by Oystein Ore, which was published by McGraw-Hill in 1948. This is particularly true of section VII which closely resembles part of Ore's chapter 6.

I. Divisibility

Reason Can Grasp What Experiment Cannot.

1. Permutations

Problem: Thirty people are to change seats in a room. How many ways can this be done?

Solution: ▪ The first person has 30 possible seats.

▪ The second has 29, giving $30 \cdot 29$ possibilities.

▪ The third has 28, giving $30 \cdot 29 \cdot 28$ possibilities.

▪ There are $30 \cdot 29 \cdot 28 \cdot 27$ possibilities at the fourth stage, and so on.

▪ At the thirtieth stage there will have been

$$30 \cdot 29 \cdot 28 \cdot 27 \cdot 26 \cdots 4 \cdot 3 \cdot 2 \cdot 1 \qquad \text{possibilities.}$$

The total number of possibilities is

$$265{,}252{,}859{,}812{,}191{,}058{,}636{,}308{,}480{,}000{,}000.$$

This number, the solution to the problem, is evenly divisible by all the numbers up to 30. This shows that mathematical reasoning goes far beyond statistical analysis: We could not obtain such a result by experimental trials.

How to Recognize Integral Factors

2. Divisibility Criteria

The character of a number is often shown through its divisors. The problem then arises of identifying the divisors of a number. In the list that follows, a way to recognize whether or not the number at the left is a divisor is shown.

2: Check that the last digit is even.

3: Add up the digits of the given number. If they have a sum that is divisible by three, then the original number is divisible by three as well. In this way one can reduce the problem to that involving a smaller number.

4: Consider the last two digits as a single number themselves. If they are divisible by four, then the original number is also.

5: The final digit must be 0 or 5.

6: Test for divisibility by both two and three.

7: [There is no convenient test.]

8: The last three digits taken as a number by themselves must be divisible by eight.

9: Find the sum of all the digits. If this sum is divisible by 9, then the original is as well.

10: There is a zero at the end.

11: Add every other digit, obtaining two sums depending on which digits are added. Subtract the smaller from the larger. The result must be divisible by eleven.

12: Check that it is divisible by both four and three.

Sample Problems

Write an integer and place it in a box so that no confusion can arise. List the integers from 2 to 12 nearby, then apply the criteria of this section to the boxed number in order to find its divisors. Cross out those numbers in the list that are *not* divisors of the boxed integer.

360| 2 3 4 5 6 X 8 9 10 XX 12

480| 2 3 4 5 6 X 8 9 10 XX 12

<u>630</u>| 2 3 X̶ 5 6 7 X̶ 9 10 X̶ X̶

<u>225</u>| X̶ 3 X̶ 5 X̶X̶X̶ 9 X̶ X̶ X̶

Sample Problem: The following number is missing its last digit:

78,543,567,92?

What digit can be put in the last place to make the number divisible by 9?

Solution: The sum of the digits is 56; so 7 must be added on.

Sample Problem: What digit can be added onto the end of the number in the previous problem in order to make it divisible by eleven?

Solution:

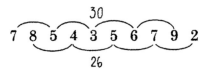

Since 30 − 26 = 4, this number itself is not divisible by 11. Adding seven to it, however, would give 37 − 26 = 11. So the last digit should be 7.

II. Congruences

Scales

3. Congruence in a Scale

The subject of congruences was called *the relationship of a number to a scale* by Gauss. As a first example, consider a scale of five.

$\lfloor 0 \rfloor$ 1 2 3 4 $\lfloor 5 \rfloor$ 6 7 8 9 $\lfloor 10 \rfloor$ 11 12 13 14 $\lfloor 15 \rfloor$ 16 17 18 19

In reference to this scale, 13 is equivalent to 8. They are similarly positioned in the scale. It is said that 13 is equivalent to 8 *modulo five*. Or, in symbols,

$$13 \equiv 8 \ (\text{mod } 5).$$

The scale of five can also be continued negatively.

-11 $\lfloor -10 \rfloor$ -9 -8 -7 -6 $\lfloor -5 \rfloor$ -4 -3 -2 -1 $\lfloor 0 \rfloor$

So we can also write $\quad -8 \equiv -3$ modulo 5.

A scale of 7 can be made in the same way.

$\lfloor 0 \rfloor$ 1 2 3 4 5 6 $\lfloor 7 \rfloor$ 8 9 10 11 12 13 $\lfloor 14 \rfloor$ 15 16 17 18 19 20 $\lfloor 21 \rfloor$

This shows us that: $\quad 13 \equiv 6 \ (\text{mod } 7);$
$$17 \equiv 10 \ (\text{mod } 7);$$
$$35 \equiv 7 \ (\text{mod } 7);$$

because the numbers on both sides of the congruence sign, $\equiv$, are similarly placed in the scale. Equivalent numbers are those that have the same remainder when divided by seven.

Computational Methods

4. The Scale of Three

It is interesting to investigate the powers of ten according to various scales.

$$10 \equiv 1 \quad (\text{mod } 3)$$
$$100 \equiv 1 \quad (\text{mod } 3)$$
$$1000 \equiv 1 \quad (\text{mod } 3)$$
$$10,000 \equiv 1 \quad (\text{mod } 3)$$

Any number can be expressed in the form

$$(\bigstar) \qquad a_0 + a_1 10 + a_2 10^2 + a_3 10^3 + a_4 10^4$$

The a's are merely the digits that make up the number when it is written with ordinary numerals.

This number is, in the scale mod 3, equivalent to:

$$a_1 + a_2 + a_3 + a_4 + a_5 = \text{the sum of its digits.}$$

The reason for this is that the various powers of ten, in ($\bigstar$) above, are all equivalent to one when taken mod 3.

5. The Scale of Seven

We shall now shift attention to the scale given by the number seven.

$$10 \equiv 3 \quad (\text{mod } 7)$$
$$100 \equiv 2 \quad (\text{mod } 7)$$
$$1000 \equiv -1 \quad (\text{mod } 7)$$
$$10,000 \equiv -3 \quad (\text{mod } 7)$$
$$100,000 \equiv -2 \quad (\text{mod } 7)$$
$$1,000,000 \equiv 1 \quad (\text{mod } 7)$$
$$10,000,000 \equiv 3 \quad (\text{mod } 7)$$

In this way we obtain a repeating sequence of remainders:

$$1, 3, 2, -1, -3, -2, 1, \text{ etc.}$$

6. The Scale of Nine

The sequence of remainders in the scale of nine is like that in the scale of three.

$$10 \equiv 1 \quad (\text{mod } 9)$$
$$100 \equiv 1 \quad (\text{mod } 9)$$
$$1000 \equiv 1 \quad (\text{mod } 9)$$
$$10{,}000 \equiv 1 \quad (\text{mod } 9)$$

As in the case of the scale given by three, the equivalent of a number may be found by forming the sum of its digits. That is, if we add the digits of a number, we obtain another, lesser number whose position in the 9's scale is exactly that of the original.

7. Testing with Multiplication

It is instructive to carry out a test of our methods by checking its effects on a multiplication. First form an arbitrary product, for example:

$$728{,}223 \times 5{,}535{,}064 = 4{,}030{,}760{,}911{,}272.$$

The first of these factors, 728,223, has 24 as the sum of its digits. So

$$728{,}223 \equiv 24 \ (\text{mod } 9).$$

Continuing this process with the number 24, we have $2 + 4 = 6$. So 728,223 is ultimately equivalent to 6 in the scale of 9. We can thus write

$$728{,}223 \equiv 6 \quad (\text{mod } 9).$$

Or equivalently, $\quad 728{,}223 \equiv -3 \ (\text{mod } 9).$

The other factor, 5,535,064, has 28 as the sum of its digits; $2 + 8 = 10$ and, to continue, $1 + 0 = 1$. So as a result we have that

$$5,535,064 \equiv 1 \ (\mathrm{mod}\ 9).$$

Thus, our two factors have the remainders -3 and 1 respectively. So their product, 4,030,760,911,272, should, if everything is correct, have a remainder equivalent to $(-3) \times (1) = -3$.

The check should give us the result that the product 4,030,760,911,272 is also equivalent to -3 (mod 9). Adding these digits:

$$4 + 0 + 3 + 0 + 7 + 6 + 0 + 9 + 1 + 1 + 2 + 7 + 2 = 42;$$

To find the remainder of 42, mod 9, we may repeat the process and add the digits once again.

$$4 + 2 = 6.$$

So the product is equivalent to 6. This is also the same, mod 9, as being equivalent to -3.

This completes the multiplicative check of our methods. A brief algebraic calculation will show the reason that this multiplication check ought to succeed. Consider two numbers to be multiplied

$$n_1 \cdot n_2.$$

Let us say that $n_1 \equiv b \ (\mathrm{mod}\ 9)$ and $n_2 \equiv d \ (\mathrm{mod}\ 9)$. That is to say that there are numbers a and c with

$$n_1 = 9a + b; \qquad\qquad n_2 = 9c + d.$$

Multiplying gives:

$$n_1 \cdot n_2 = (9a + b)(9c + d) = 81ac + 9bc + 9ad + bd = 9f + bd,$$

where $f = 9ac + bc + ad$.

As a result $n_1 \cdot n_2 \equiv bd \ (\mathrm{mod}\ 9)$. This justifies our use of the multiplication test.

8. A Note on Reversed Digits

The difference between two numbers written with reversed digits is divisible by nine. For example,

$$843,965,238,743$$
$$\underline{347,832,569,348}$$
$$496,132,669,395$$

The sum of the digits of this last number is 63, so it is divisible by 9. The general principle here can be seen by beginning with the two numbers n_1 and n_2; giving names to their digits we have:

$$n_1 = a_0 + a_1 10 + a_2 10^2 + a_3 10^3 + a_4 10^4 ;$$
$$n_2 = a_4 + a_3 10 + a_2 10^2 + a_1 10^3 + a_0 10^4 .$$

Subtracting, we obtain the difference, $\qquad n_1 - n_2 =$

$$(a_0 - a_4) + (a_1 - a_3)10 + (a_2 - a_2)10^2 + (a_3 - a_1)10^3 + (a_4 - a_0)10^4$$

$$\equiv (a_0 - a_4) + (a_1 - a_3) + (a_2 - a_2) + (a_3 - a_1) + (a_4 - a_0) \quad (\text{mod } 9)$$

$$\equiv 0 \ (\text{mod } 9).$$

9. The 99 Scale

The scale given by 99 generates a pattern in which digits repeat in cycles of two:

$$1 \equiv 1 \quad (\text{mod } 99)$$
$$10 \equiv 10 \quad (\text{mod } 99)$$
$$100 \equiv 1 \quad (\text{mod } 99)$$
$$1,000 \equiv 10 \quad (\text{mod } 99)$$
$$10,000 \equiv 1 \quad (\text{mod } 99) , \text{ and so on.}$$

For this reason, testing for equivalence in the scale of 99 is done by considering the digits of a number in groups of two.

For example, to determine the remainder when the number 44,910,558 is divided by 99, we may group the digits in pairs as follows:

$$44 \mid 91 \mid 05 \mid 58.$$

Adding these numbers $44 + 91 + 05 + 58 = 198$, we obtain 198. Carrying out the process again, 198 is broken into pairs of digits, starting as usual from the right hand side:

$$01 \mid 98.$$

Then the pairs are added as two digit numbers, giving $1 + 98 = 99$. Thus, the original number 44,910,558 is divisible by 99.

10. The 999 Scale

Next let us turn to the scale given by 999:

$$1 \equiv 1 \quad (\text{mod } 999)$$
$$10 \equiv 10 \quad (\text{mod } 999)$$
$$100 \equiv 100 \quad (\text{mod } 999)$$
$$1,000 \equiv 1 \quad (\text{mod } 999)$$
$$10,000 \equiv 10 \quad (\text{mod } 999)$$
$$100,000 \equiv 100 \quad (\text{mod } 999)$$
$$1,000,000 \equiv 1 \quad (\text{mod } 999)$$

This means that we can test for equivalence in the scale of 999 by considering the digits of a number in groups of three. As an example, consider the arbitrary multiplication that was used in section 7:

(☆) $728,223 \times 5,535,064 = 4,030,760,911,272.$

We carry out the test for 999 on the first of these factors: 728,223.

$$\begin{array}{r} 223 \\ + \ 728 \\ \hline 951 \end{array} \equiv -48 \ (\text{mod } 999)$$

The digits have been added in groups of three. We obtain the result that the first factor in the product is equivalent to −48, modulo 999. Now do the same for the other factor, 5,535,064.

$$064$$
$$535$$
$$5$$
$$\overline{}$$
$$604 \equiv -395 \ (\text{mod } 999)$$

Finally, the product 4,030,760,911,272 is tested by putting its digits in groups of three as well.

$$272 + 911 + 760 + 030 + 4 = 1977$$

Since this is larger than 999, we carry out the process again, still adding the digits in groups of three.

$$977$$
$$1$$
$$\overline{}$$
$$978 \equiv -21 \ (\text{mod } 999)$$

So, the two factors appearing in (✩) have been reduced to their equivalents mod 999, namely −48 and −395 respectively: and, in addition, their product in (✩) is equivalent to −21, mod 999.

To carry out the check on our calculations, multiply the values −48 and −395 that we have obtained as equivalents of the two factors.

Multiplying: $\qquad$ −395
$\qquad\qquad\qquad$ −48
$\qquad\qquad\qquad \overline{}$
$\qquad\qquad\qquad$ 18,960

Now reduce the product 18,960 by grouping the digits in groups of three as we do in the scale of 999.
Adding the groups,

$$960 + 18 = 978 \equiv -21 \ (\text{mod } 999)$$

This is what we expected to happen if everything were to be correct because the product (✶) has already given –21.

To illustrate the computation again, let us use 604 in the product instead of –395. These have already been shown to be equivalent, mod 999, in the computation immediately following the original product (✶), and we should expect them to give equivalent results.

$$(-48) \times (604) = -28{,}992.$$

Repeating the reduction of this last number, we carry out the computation by grouping the digits in triples as before:

$$992 + 28 = 1{,}020.$$

Since –28,992 is negative, it remains negative after reduction. So the product is equivalent to –1,020 (mod 999).

Now we reduce again; this time the triples appearing in the product are "020" and "1".

$$
\begin{array}{r}
20 \\
1 \\
\hline
-\ 21
\end{array}
$$

Every way of proceeding leads us to the observation that multiplication is respected by the use of equivalences in the scale of 999.

11. The Scale of Eleven

Next, let us form the scale given by eleven.

$$
\begin{aligned}
1 &\equiv 1 \quad (\mathrm{mod}\ 11) \\
10 &\equiv -1 \quad (\mathrm{mod}\ 11) \\
100 &\equiv 1 \quad (\mathrm{mod}\ 11) \\
1000 &\equiv -1 \quad (\mathrm{mod}\ 11)
\end{aligned}
$$

So the corresponding sequence is 1,–1, 1,–1, 1,–1. This means that a number of the form

$$a_0 + a_1 10 + a_2 10^2 + a_3 10^3 + a_4 10^4$$

will be equivalent to $a_0 - a_1 + a_2 - a_3 + a_4$ (mod 11).

12. The 101 Scale

$$1 \equiv 1 \quad (\text{mod } 101)$$
$$10 \equiv 10 \quad (\text{mod } 101)$$
$$100 \equiv -1 \quad (\text{mod } 101)$$
$$1{,}000 \equiv -10 \quad (\text{mod } 101)$$
$$10{,}000 \equiv 1 \quad (\text{mod } 101)$$
$$100{,}000 \equiv 10 \quad (\text{mod } 101)$$

We may calculate equivalents in the scale of 101 by alternately adding and subtracting, as in the case of 11, but this time using the digits in groups of two.

Next, we shall test this technique to see if it is indeed preserved under multiplication. Begin with the same product that was called (☆) in section 10.

$$728{,}223 \times 5{,}535{,}064 = 4{,}030{,}760{,}911{,}272.$$

Now break down each of these three numbers in groups of two digits that will be alternately added and subtracted according to the method for testing divisibility by 101.

72| 82|23

5| 53| 50|64

4| 03| 07| 60| 91| 12| 72

The first of these is handled in the computation below at the left; the second is in the middle, and the third at the right. Every other one of the two digit, breakdown numbers are added and then the alternate ones are subtracted from the resulting sum.

```
    23              64              72
    72              53              91
   ───             ───              07
    95        .    117               4
   -82             -55             ───
   ───             ───             174
    13              62             -12
                                   -60
                                    -3
                                   ───
                                    99
```

$99 \equiv -2$ (mod 101), so we should expect our check to verify that 13×62 also yields a -2. Now $13 \times 62 = 806$. Grouping the digits in pairs gives 8 | 06, which reduces to $06 - 8 = -2$. This serves to check the method.

13. The 1001 Scale

$$1 \equiv 1 \quad \text{(mod 1001)}$$
$$10 \equiv 10 \quad \text{(mod 1001)}$$
$$100 \equiv 100 \quad \text{(mod 1001)}$$
$$1,000 \equiv -1 \quad \text{(mod 1001)}$$
$$10,000 \equiv -10 \quad \text{(mod 1001)}$$
$$100,000 \equiv -100 \quad \text{(mod 1001)}$$

The calculation of equivalents in the scale of 1001 is carried out by alternately adding and subtracting as before, but now the digits are grouped in triples. For example, choose the number 4,690,009,324. Having separated the digits in blocks of three, we may alternately add and subtract them as shown below.

$$-4$$
$$690$$
$$-009$$
$$324$$

$$\overline{}$$

$$1001$$

This shows that $4,690,009,324$ is a multiple of 1001.

As before we may check the method by multiplying. Begin with an arbitrary product. We may as well use the same numbers that have been used with the other scales for the purpose of checking.

$$728,223 \times 5,535,064 = 4,030,760,911,272.$$

To carry out the check, obtain the remainders mod 1001 using the method of alternately adding and subtracting the blocks of three.

223	64	272
−728	5	760
	−535	4
−505		
	−466	1036
		−941
		95

So we shall hope to find that $(-505) \times (-466) =$ 505×466 also gives 95. Multiplying, we find that the product of 505 and 466 is $235,330$. As before we must break these into triples $235 \mid 330$, and subtract. The result checks our method.

14. Checking with Division

We may also check our methods by using division, not just multiplication as has been done previously. Begin with an arbitrary division. Since we are concerned only with whole numbers, there will be a remainder.

$$\frac{76,638,123}{37,547} = 2041 \quad R \ 4696$$

First find the equivalent values mod 9, by adding the digits.

$$37,547 \equiv -1 \quad (\text{mod } 9)$$
$$2,041 \equiv -2 \quad (\text{mod } 9)$$
$$4,696 \equiv -2 \quad (\text{mod } 9)$$
$$76,638,123 \equiv \ 0 \quad (\text{mod } 9)$$

Now we can check the congruences $(-1)(-2) + (-2) = 0$. Which is the result that we would expect. In this example the calculation gave 0 exactly, but in general one should only expect that there be an equivalence rather than an actual equality. This can be seen by testing the same division in a different scale. In the scale 11 we alternately added and subtracted the digits of a number. This gives

$$37,547 \equiv \ 4 \quad (\text{mod } 11)$$
$$2,041 \equiv -5 \quad (\text{mod } 11)$$
$$4,696 \equiv -1 \quad (\text{mod } 11)$$
$$76,638,123 \equiv \ 1 \quad (\text{mod } 11).$$

This test gives $4 \times (-5) + (-1) = -21 \equiv 1 \ (\text{mod } 11)$

Before examining the reason for this let us test another scale on the same division. The scale of 99 will do nicely. In this scale the digits of the number to be evaluated were grouped in blocks of two.

$$37,547 \equiv 26 \quad (\text{mod } 99)$$
$$2,041 \equiv 61 \quad (\text{mod } 99)$$
$$4,696 \equiv 43 \quad (\text{mod } 99)$$
$$76,638,123 \equiv 45 \quad (\text{mod } 99)$$

So to carry out the check we shall expect to find that $26 \times 61 + 43$ is equivalent to 45 too. Multiplying, we find that $26 \times 61 = 1586$. Breaking down the number into blocks of length two, we obtain 86 + 15= 101 ≡ 2 (mod 99). The remainder of 4696 has been found equivalent to 43. So finally, 2 + 43 = 45 as expected.

To justify this test an algebraic proof is needed. Working with remainders mod 9, the original division involving a divisor, quotient, and remainder can be expressed as

$$(n_1 \cdot 9 + n_2) \cdot (n_3 \cdot 9 + n_4) + (n_5 \cdot 9 + n_6) = (n_7 \cdot 9 + n_8) .$$

Expanding gives

$$n_1 n_3 \cdot 9^2 + n_2 n_3 \cdot 9 + n_1 n_4 \cdot 9 + n_2 n_4 + n_5 \cdot 9 + n_6$$

This is equivalent mod 9 to $n_2 n_4 + n_6$; meanwhile the right side is equivalent to n_8, so it must be that

$$n_2 n_4 + n_6 \equiv n_8.$$

Alternatively, if we reduce the expressions mod 9 in the first place, rather than waiting, we obtain

$$n_1 \cdot 9 + n_2 \equiv n_2 \text{ (mod 9)}; \qquad n_5 \cdot 9 + n_6 = n_6 \text{ (mod 9)};$$
$$n_3 \cdot 9 + n_4 \equiv n_4 \text{ (mod 9)}; \qquad n_7 \cdot 9 + n_8 = n_8 \text{ (mod 9)}.$$

So the reduced values satisfy $n_2 n_4 + n_6 \equiv n_8$.

Here are some additional examples of the preservation of arithmetical operations under congruence. To begin with, another division with remainder:

$$24,726,928,309 \div 6927 = 3,569,644 \text{ R } 4321.$$

We may reduce this last expression, mod 9, by adding the digits of each number and obtain

$$(-2) \div (-3) = 1 \quad \text{R } 1.$$

This checks, as shown by the computation

$$(1)(-3) + 1 \equiv -2 \text{ (mod 9)}$$

Now, as another example, let us consider a square number.
$$(41,587)^2 = 1,729,478,569$$

When it has been reduced mod 9 these become

$$(-2)^2 \text{ and } 13.$$

Yielding $\qquad\qquad 4 \equiv 4 \pmod 9.$

Additions and subtractions are preserved too, as illustrated by the following examples.

468,322,159	→	4		
248,564,235	→	3	468,322,159 → 4
118,300,247	→	-1	-248,564,235 → -3
835,186,641	→	6	219,757,924 → 1

III. Aliquot Parts

Numerical Friendship

15. Amicable Pairs

The first pair of amicable numbers were found by Pythagoras. They are

$$\boxed{220 \text{ and } 284}$$

To see what this means, reduce these numbers by removing prime divisors. For example, 2 divides 220, so we may remove a factor of 2 and leave 110. The factors that have been successively removed are listed to the right of the numbers from which they shall be taken. The quotient is put on the next line. Then we start again and look for another prime factor. Eventually we arrive at 1 and stop.

220	2		284	2
110	2		142	2
55	5		71	71
11	11		1	
1				

This computation can provide us with a list of *all* the divisors of 220 and 284, not merely the primes. We must multiply the prime factors in their various combinations to obtain the other divisors. We get 1, 2, 4, 5, 10, 11, 20, 22, 44, 55, and 110 as divisors of 220, and 1, 2, 4, 71, 142 for 284. These divisors are said to be the *aliquot parts* of 220 and 284. Now, adding the aliquot parts together we obtain

for 220

$$1 + 2 + 4 + 5 + 10 + 11 + 20 + 22 + 44 + 55 + 110 = 284$$

for 284

$$1 + 2 + 4 + 71 + 142 = 220$$

Amicable numbers are pairs of numbers each in which each is the sum of the aliquot parts of the other. In addition, 220 is said to be an *abundant* number because the sum of its aliquot parts is greater than 220 itself. On the other hand, 284 is said to be *deficient*: that is, the sum of the aliquot parts of 284 is less than 284.

It was a very long time before any other amicable pairs were found. Approximately 2000 years later the second pair was discovered by Ternat in 1636. In 1638 Descartes found a third pair. In 1846 Nicolo Paganini found the fourth pair of amicable numbers when he was 16 years old.

2)	17,296	18,416
3)	9,363,548	9,437,056
4)	1,184	1,210

To illustrate these definitions again, consider the last of these boxed pairs. First, reduce the numbers by removing their prime factors.

$$1184 \big\lfloor\ 2$$

$$592 \big\lfloor\ 2$$

$$296 \big\lfloor\ 2$$

$$148 \big\lfloor\ 2$$

$$74 \big\lfloor\ 2$$

$$37 \big\lfloor\ 37$$

$$1$$

Next multiply the prime factors in their various combinations to obtain all the factors.

Aliquot Parts: 1, 2, 4, 8, 16, 32, 37, 74, 148, 296, 592

Sum of the Aliquot Parts: 1210

So, the aliquot parts of 1184 add to 1210. Now it must be checked that, conversely, the aliquot parts of 1210 add to 1184.

$$1210 \ |\ 2$$
$$605 \ |\ 5$$
$$121 \ |\ 11$$
$$11 \ |\ 11$$
$$1$$

Multiplying these in their various combinations gives us still more factors 10, 22 and so on.

Aliquot Parts: 1, 2, 5, 10, 11, 22, 55, 110, 121, 242, 605.

Sum of the Aliquot Parts: 1184

16. Perfection

Analyzing the number 28, the prime factors are found as follows:

$$28 \ |\ 2$$
$$14 \ |\ 2$$
$$7 \ |\ 7$$
$$1$$

By multiplying these factors in various combinations, we obtain the aliquot parts: 1, 2, 4, 7, 14. These have 28 itself as their sum. One can say that multiplication and addition are perfectly balanced in the number 28. Such numbers are said to be *perfect*.

17. Deficiency and Abundance

When the sum of the aliquot parts falls short of the number itself, it is said to be *deficient*; when the aliquot parts exceed the number in sum, then the number is *abundant*. The following table shows the abundance or deficiency of the first few natural numbers. The aliquot parts of the number appear in the second column and the sum of the aliquot parts in the third.

	parts	sum	type
2	1	1	deficient
3	1	1	deficient
4	1, 2	3	deficient
5	1	1	deficient
6	1, 2, 3	6	perfect
7	1	1	deficient
8	1, 2, 4	7	deficient
9	1, 3	4	deficient
10	1, 2, 5	8	deficient
11	1	1	deficient
12	1, 2, 3, 4, 6	16	abundant
13	1	1	deficient
14	1, 2, 7	10	deficient
15	1, 3, 5	9	deficient
16	1, 2, 4, 8	15	deficient
17	1	1	deficient
18	1, 2, 3, 6, 9	21	abundant
19	1	1	deficient
20	1, 2, 4, 5, 10	22	abundant
21	1, 3, 7	11	deficient
22	1, 2, 11	14	deficient
23	1	1	deficient
24	1,2,3,4,6,8,12	36	abundant
25	1, 5	6	deficient
26	1, 2, 1 3	16	deficient
27	1, 3, 9	13	deficient
28	1, 2, 4, 7, 14	28	perfect
29	1	1	deficient
30	1,2,3,5,6,10,15	42	abundant

The deficient numbers, those that are greater than the sum of their aliquot parts, are initially in the majority. The first abundant numbers are 12, 18, 20 and 24.

The number 60 factors into $2^2 \cdot 3 \cdot 5$, so, multiplying these prime factors in various combinations, we can obtain the aliquot parts of 60: 1, 2, 3, 4, 5, 6, 10, 12, 15, 20 and 30. These add to 108. The ratio $^{108}\!\!/_{60} = 1.8$ gives a measure of the abundance of the number 60.

An even more abundant number is $120 = 2^3 \cdot 3 \cdot 5$. The aliquot parts are: 1, 2, 3, 4, 5, 6, 8, 10, 12, 15, 20, 24, 30, 40, and 60. Their sum is 240, giving a *ratio of abundance* of $^{240}\!\!/_{120} = 2$.

The number 360 is very highly divisible. It factors into $2^3 \cdot 3^2 \cdot 5$. This gives an extensive list of aliquot parts 1, 2, 3, 4, 5, 6, 8, 9, 10, 12, 15, 18, 20, 24, 30, 36, 40, 45, 60, 72, 90, 120, and 180. The sum of the aliquot parts is 810. The ratio of abundance is $^{810}\!\!/_{360} = 2.25$.

Since 360 is such an abundant number it was widely used in ancient measuring systems because it allowed divisions to be carried out without any fractions appearing.

18. The Smallest Perfect Numbers

Twelve perfect numbers have been found, the first four of which are 6, 28, 496, and 8128. To verify that these numbers are indeed perfect, factor them into primes, obtain the aliquot parts, and then find the sum of these aliquot parts.

$$6 \mid 2$$
$$3 \mid 3$$
$$1 \mid$$

Having thus factored 6, we obtain the factors 2 and 3. The aliquot parts of 6 are therefore 1, 2, and 3. Thus 6 is perfect: Certainly $1 + 2 + 3 = 6$.

Factoring 28, the next perfect number gives us

$$28 \lfloor 2$$

$$14 \lfloor 2$$

$$7 \lfloor 7$$

$$1 \lfloor$$

Aliquot Parts: 1, 2, 4, 7, 14.　　　1 + 2 + 4 + 7 + 14 = 28.

The factorization of 496 proceeds as follows:

$$496 \lfloor 2$$

$$248 \lfloor 2$$

$$124 \lfloor 2$$

$$62 \lfloor 2$$

$$31 \lfloor 31$$

$$1 \lfloor$$

This gives $2^4 \cdot 31$ as the prime factors of 496.

The aliquot parts then are 1, 2, 4, 8, 16, 31, 62, 124, and 248. It is easily checked that their sum is 496.

The last example is 8128.

$8128 \lfloor 2$	$508 \lfloor 2$
$4064 \lfloor 2$	$254 \lfloor 2$
$2032 \lfloor 2$	$127 \lfloor 127$
$1016 \lfloor 2$	$1 \quad \lfloor$

$$\text{Thus } 8128 = 2^6 \cdot 127.$$

The aliquot parts of 8128 are 1, 2, 4, 8, 16, 32, 64, 127, 254, 508, 1016, 2032, and 4064. Again, their sum is the number itself, 8128.

19. Euclid's Approach to Perfect Numbers

We have seen that $6 = 2 \cdot 3,$

$$28 = 2 \cdot 2 \cdot 7,$$

$$496 = 2 \cdot 2 \cdot 2 \cdot 2 \cdot 31,$$

and $8128 = 2 \cdot 2 \cdot 2 \cdot 2 \cdot 2 \cdot 2 \cdot 127.$

There is a pattern involving these particular primes 3, 7, 31, and 127 that is revealed by writing a list of powers of two.

In the list that follows, powers of two are written in a vertical column. To the right of each entry let us record the integer that is one less than it.

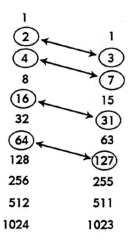

The right-hand column consists of numbers that are one less than the power of two that appears at the left. When primes appear they have been circled. Primes of this kind, that are one less than a power of two, are called *Mersenne primes*.

The primes are associated with a power of two in our factorization of the perfect numbers. This associated power of two has been linked to its corresponding prime by an arrow.

So these perfect numbers can be seen to have the form

$$N = 2^n (2^{n+1} - 1)$$

for some number n having the property that $2^{n+1}-1$ is a prime number. In fact, our four previous examples of perfect numbers suggest that we search for perfect numbers among those numbers for which n itself is prime.

Turning our attention to prime numbers having the form $2^{p+1} - 1$, where p is prime (the Mersenne primes), it is known that for $p = 127$ the number $2^{p+1} - 1$ is prime. This number,

$$170,141,183,460,469,231,731,687,303,884,105,727$$

is the largest prime of this kind among the numbers having $p \leq 257$. It is unknown whether or not an odd, perfect number can exist.

20. Multiple Perfect Numbers

Let us begin by reviewing the calculation of aliquot parts, deficiency, and abundance.

Starting with 1387, we remove prime factors as shown.

$$1387 \; \big|\; 19$$

$$73 \; \big|\; 73 \qquad \text{(Factor tables may be used to speed the work.)}$$

$$1 \; \big|$$

Since there are only two prime factors it is very easy to list the aliquot parts and then find their sum, $1 + 19 + 73 = 93$. So 1387 is seen to be deficient.

To pick another example, let us analyze 20,677. First, it must be factored.

$$20,677 \mid 23$$

$$899 \mid 29$$

$$31 \mid 31$$

So we have that $20,677 = 23 \cdot 29 \cdot 31$. The aliquot parts are

$$
\begin{array}{l}
1 \\
23 \\
29 \\
31 \\
667 = 23 \cdot 29 \\
713 = 23 \cdot 31 \\
899 = 29 \cdot 31 \\
\hline
\end{array}
$$

sum: 2363

So 20,677 is deficient.

Now let us turn to a more interesting example: 672. We shall proceed as in the previous examples. First, the factorization:

$$672 \mid 2$$

$$336 \mid 2$$

$$168 \mid 2$$

$$84 \mid 2$$

$$42 \mid 2$$

$$21 \mid 3$$

$$7 \mid 7$$

$$1 \mid$$

This gives the factors $672 = 2^5 \cdot 3 \cdot 7$. From these we set about finding the aliquot parts. As usual, the prime factors must be multiplied in their various combinations to obtain 1, 2, 3, 4, 6, 7, 8, 12, 14, 16, 21, 24, 28, 32, 42, 48, 56, 84, 96, 112, 168, 224, and 336.

Their sum is 1344. The ratio $\frac{1344}{672} = 2$, is in this case an integer. We may call 1344 a multiple perfect number; its abundance is 2.

The number 523,776 also has an abundance of 2. It factors into $2^9 \cdot 3 \cdot 11 \cdot 31$, as can be verified by considering the products that follow:

$$523,776 = 261,888 \cdot 2;$$
$$261,888 = 130,944 \cdot 2;$$
$$130,944 = \ 65,472 \cdot 2$$
$$65,472 = 32,736 \cdot 2$$
$$32,736 = 16,368 \cdot 2$$
$$16,368 = 8,148 \cdot 2;$$
$$8,148 = 4,092 \cdot 2;$$
$$4,092 = 2,046 \cdot 2$$
$$2,046 = 1,023 \cdot 2;$$
$$1,023 = 341 \cdot 3;$$
$$341 = 31 \cdot 11$$

The aliquot parts of 523,776 are given below in pairs of factors.

2	261,888	44	11,904	192	2,728
3	174,592	48	10,912	248	2,112
4	130,944	62	8,448	256	2,046
6	87,296	64	8,184	264	1,984
8	65,472	66	7,936	341	1,536
11	47,616	88	5,952	352	1,488
12	43,648	93	5,632	372	1,408
16	32,736	96	5,456	384	1,364
22	23,808	124	4,224	496	1,056
24	21,824	128	4,092	512	1,023
31	16,896	132	3,968	528	992
32	16,368	176	2,976	682	768
33	15,872	186	2,816	704	744

The total of all of these divisors (the number 1 must also be included in the sum) is 1,047,552 or twice 523,776 itself.

Prime Numbers

21. The Sieve

The sieve method is the only complete method for obtaining all of the prime numbers. It cannot be captured in a formula that could then be used to generate the primes.

In the table that follows there are boxes placed around the multiples of 2, 3, 5, and 7.

1	2	3	4	5	6	7	8	9	10
11	12	13	14	15	16	17	18	19	20
21	22	23	24	25	26	27	28	29	30
31	32	33	34	35	36	37	38	39	40
41	42	43	44	45	46	47	48	49	50
51	52	53	54	55	56	57	58	59	60
61	62	63	64	65	66	67	68	69	70
71	72	73	74	75	76	77	78	79	80
81	82	83	84	85	86	87	88	89	90
91	92	93	94	95	96	97	98	99	100

The multiples of 2 were obtained by boxing out all the numbers that fall in alternate columns. The multiples of 5 were found in the columns headed by either 5 or 10. Finally, the multiples of 3 were found on slanting diagonals in the table; these diagonals are rising as one goes from the left to the right. For example, beginning at 81, there is a diagonal containing 81, 72, 63, etc.

Next the multiples of 7 need to be found. Since they do not have a convenient pattern in the table it is necessary to find them in succession 7^2, $7 \cdot 11$, and $7 \cdot 13$.

When this has been done the numbers that have not been boxed will be the primes less than one hundred. There are twenty-four of them, not counting 1, which was not boxed out in the table even though it is not regarded as a prime. They are:

2, 3, 5, 7, 11, 13, 17, 19, 23, 29, 31, 37, 41

43, 47, 53, 59, 61, 67 71, 73, 79, 83, 89, 97.

When primes are as closely paired as possible, that is they are separated by two, they are said to be *twin primes*. The largest twin primes less than 100 are the pair 71 and 73.

It would be possible to carry out this process by hand in order to list all the primes less than one billion.

22. Counting Prime Numbers

Relying on lists of primes that have been found, let us count the number of primes to be found in various intervals. We shall include 1 in the count as though it were a prime.

0 – 100	25		1,000,000 – 1,000,100	6
100 – 200	21		1,000,100 – 1,000,200	10
200 – 300	16		1,000,200 – 1,000,300	8
300 – 400	16		1,000,300 – 1,000,400	8
400 – 500	17		1,000,400 – 1,000,500	7
500 – 600	14		1,000,500 – 1,000,600	7
600 – 700	16		1,000,600 – 1,000,700	10
700 – 800	14		1,000,700 – 1,000,800	5
800 – 900	15		1,000,800 – 1,000,900	6
900 – 1000	14		1,000,900 – 1,001,000	8
TOTAL	168		TOTAL	75

Let $\Pi(x)$ be the number of primes less than x (we shall add 1 to the count). Then

$$\Pi(10) \quad = \quad 4;$$
$$\Pi(100) \quad = \quad 25;$$
$$\Pi(1000) \quad = 168.$$

It is plausible that

$$\lim_{x \to \infty} \prod(x) = \frac{x}{\ln x}$$

To see that this makes sense, we may consider instead the values of the common logarithm

$$\frac{x}{\log x}.$$

When $x = 10$, $\dfrac{x}{\log x} = \dfrac{10}{1} = 10.$

When $x = 100$, this takes the value $\frac{100}{2} = 50$; and when $x = 1000$, it would estimate that there were 333 primes

23. Avoiding Primes

It is possible to pick intervals of any given length that contain no primes. Begin by looking at numbers of the form

$$x = (n + 1)!$$

in order to find such an interval of length n. Certainly x would not be prime; but $x + 1$ could be prime, although it is doubtful. It is certain, however, that $x + 2$ is divisible by 2, since 2 divides x; $x + 3$ is divisible by 3 since 3 divides x also; $x + 4$ is divisible by 4 and so on.

For example, $5! = 120$. Its successor, 121, is a doubtful prospect for being a prime number and, in fact, is 11^2. Two divides 122; three divides 123; four divides 124; and five divides 125. So we have found four successive numbers, 122, 123, 124, and 125, that are not prime. In this instance 121 is a fifth.

To find six consecutive numbers that are not prime consider $7! = 5040$. The successor 5041 is questionable, but $5042 = 5040 + 2$ is divisible by 2; $5043 = 5040 + 3$ is divisible by 3; $5044 = 5040 + 4$ is divisible by 4; and so on.

IV. Decimal Developments

Repeating Patterns in Decimal Expansions

24. Terminating and Repeating Expansions

By dividing we obtain:

$$\frac{1}{7} = 0.\underline{142\ 857}\ 142\ 857 \cdots .$$

The same six digits appear cyclically in the related quotients:

$$\frac{2}{7} = 0.285\ 7\underline{14\ 285\ 7}14 \cdots$$

$$\frac{3}{7} = 0.428\ 57\underline{1\ 428\ 571} \cdots$$

$$\frac{4}{7} = 0.57\underline{1\ 428\ 571}\ 428 \cdots$$

$$\frac{5}{7} = 0.7\underline{14\ 285\ 7}14\ 285 \cdots$$

$$\frac{6}{7} = 0.857\ \underline{142\ 857}\ 142 \cdots$$

When we obtain quotients denominated with 13 there are two different sequences of digits that appear cyclically in the products. In the list that follows one of the sequences has been underlined and the other written boldface.

$$\frac{1}{13} = 0.076\ \underline{923\ 076\ 923} \cdots$$

$$\frac{2}{13} = 0.\mathbf{153\ 846}\ 153\ 846 \cdots$$

$$\frac{3}{13} = 0.\underline{230\ 769}\ 230\ 769 \cdots$$

$$\frac{4}{13} = 0.307\ 692\ \underline{307\ 692} \cdots$$

$$\frac{5}{13} = 0.384\ 615\ \mathbf{384\ 615} \cdots$$

$$\frac{6}{13} = 0\ 461\ \mathbf{538}\ 461\ 538 \cdots$$

$$\frac{7}{13} = 0.538\ 461\ \mathbf{538\ 461} \cdots$$

$$\frac{8}{13} = 0.615\ \mathbf{384\ 615}\ 384 \cdots$$

$$\frac{9}{13} = 0.69\underline{2\ 307\ 692}\ 307 \cdots$$

$$\frac{10}{13} = 0.769\ \underline{230\ 769}\ 230 \cdots$$

$$\frac{11}{13} = 0.846\ \mathbf{153\ 846}\ 153 \cdots$$

$$\frac{12}{13} = 0.9\underline{23\ 076\ 923}\ 076 \cdots$$

Of course there are fractions whose decimal expansions terminate without stretching to infinity. The table below shows the decimal expansions of the first few reciprocals of integers.

$\frac{1}{2}$	0.5	$\frac{1}{7}$	0.142 857$\cdots$	$\frac{1}{12}$	0.083 333$\cdots$
$\frac{1}{3}$	0.333 333$\cdots$	$\frac{1}{8}$	0.125	$\frac{1}{13}$	0.076 923$\cdots$
$\frac{1}{4}$	0.25	$\frac{1}{9}$	0.111 111$\cdots$	$\frac{1}{14}$	0.071 428$\cdots$
$\frac{1}{5}$	0.2	$\frac{1}{10}$	0.1	$\frac{1}{15}$	0.066 666$\cdots$
$\frac{1}{6}$	0.166 666$\cdots$	$\frac{1}{11}$	0.090 909$\cdots$		

How can we determine which fractions will have an infinite expansion? The answer is that the expansion will halt for those fractions which can be multiplied by a power of ten to produce an integer. For example, 0.123456789 is a terminating decimal and when it is multiplied by 10^7 gives the product 1,234,567. A number of the form

$$\frac{1}{2^n 5^m}$$

can always be multiplied by a power of 10 to obtain an integer. To see this, let us consider the case in which $n > m$ first, and multiply the fraction by 10^n to obtain an integer:

$$\frac{10^n}{2^n \cdot 5^m} = \frac{2^n \cdot 5^n}{2^n \cdot 5^m} = 5^{n-m}.$$

When $m \geq n$ one can multiply by 10^m instead to obtain an integer. This way it will be a power of two.

Let us consider the number 320 for example. It factors as $320 = 2^6 \cdot 5$. So, when we multiply $\frac{1}{320}$ by 10^6 it will become an integer. This means that the decimal expansion will go on for six places before it terminates.

Another example is $1280 = 2^8 \cdot 5$; this number will terminate only after eight decimal places.

25. Cyclic Permutations Mod Seven

A permutation is an action, "mutate," that changes something, "per," that is, through and through. In examining these decimal expansions we find a cyclic permutation of the digits:

$$\frac{1}{7} = 0.142\ 857\ \cdots$$
$$\frac{3}{7} = 0.428\ 571\ \cdots$$
$$\frac{2}{7} = 0.285\ 714\ \cdots$$
$$\frac{6}{7} = 0.857\ 142\ \cdots$$
$$\frac{4}{7} = 0.571\ 428\ \cdots$$
$$\frac{5}{7} = 0.714\ 285\ \cdots$$

To better understand this pattern, start by dividing $\frac{1}{7}$ to obtain $0.142\ 857\ \cdots$; it would do equally well to divide $\frac{1,000,000}{7}$ and obtain $142,857\cdots$. By the same line of thought, the division $\frac{10}{7}$ would give $1.428,571\cdots$.

Now, $10 \equiv 3 \pmod{7}$, so the fractional part of $\frac{3}{7}$ is the same as that of $\frac{10}{7}$, that is to say $1.478571\cdots$. This is the second in the list of cyclic permutations given above.

Continuing in this way $\frac{100}{7} = 14.785\ 714\ \cdots$; we have that $100 \equiv 2 \pmod{7}$ so that the fractional part of $\frac{100}{7}$ and that of $\frac{2}{7}$ are equal. Both are $.287\ 714\ \cdots$. Similarly:

$$\frac{1000}{7} = 142.857\ 1428\ \cdots;\ 10^3 \equiv 6 \pmod{7}\ ;(\text{ compare with } \frac{6}{7}\ ):$$

$$\frac{10,000}{7} = 1428.571\ 428\ \cdots;\ 10^4 \equiv 4 \pmod{7}\ ;\ (\text{compare with } \frac{4}{7}):$$

$$\frac{100,000}{7} = 14285.714\ 28\ \cdots;\ 10^5 \equiv 5 \pmod{7}\ ;\ (\text{compare with } \frac{5}{7}).$$

From this we see that the sequence digits follows upon the congruences of the powers of 10.

26. Cyclic Permutations Mod 13

There are six digits in the cycle obtained by dividing $\frac{1}{13}$; they also appear in the expansion of $\frac{1,000,000}{13}$ which gives the sequence $76,923.07\cdots$. The same digits appear under cyclic permutations in the fractions

$$\frac{1}{13} = .076\ 923\ \cdots\ ;$$

$$\frac{10}{13} = .769\ 230\ \cdots\ ;$$

$$\frac{9}{13} = .692\ 307\ \cdots\ ;$$

$$\frac{12}{13} = .923\,076\,\cdots\,;$$

$$\frac{3}{13} = .230\,769\,\cdots\,;$$

$$\frac{4}{13} = .307\,692\,\cdots\,.$$

Again we shall compare these expansions with the congruences mod 13.

$$2 \equiv 2 \ (\text{mod } 13); \qquad \frac{2}{13} = 0.153\,846\,\cdots\,;$$

$$20 \equiv 7 \ (\text{mod } 13); \qquad \frac{7}{13} = 0.538\,461\,\cdots\,;$$

$$200 \equiv 5 \ (\text{mod } 13); \qquad \frac{5}{13} = 0.384\,615\,\cdots\,;$$

$$2{,}000 \equiv 11 \ (\text{mod } 13); \qquad \frac{11}{13} = 0.846\,153\,\cdots\,;$$

$$20{,}000 \equiv 6 \ (\text{mod } 13); \qquad \frac{6}{13} = 0.461\,538\,\cdots\,;$$

$$200{,}000 \equiv 8 \ (\text{mod } 13). \qquad \frac{8}{13} = 0.615\,384\,\cdots\,.$$

27. Additional Examples

The reciprocal of 11 has a decimal expansion whose digits repeat in blocks of two:

$$\frac{1}{11} = 0.090\,909\,\cdots\,;$$

likewise there is a cycle of length that can be found in the congruences

$$1 \equiv 1 \quad (\text{mod } 11);$$
$$10 \equiv 10 \ (\text{mod } 11);$$
$$100 \equiv 1 \quad (\text{mod } 11).$$

The fraction $\dfrac{1}{3} = 0.333 \cdots$ can be compared with the constant sequence of remainders given by

$$1 \equiv 1 \ (\text{mod } 3);$$
$$10 \equiv 1 \ (\text{mod } 3);$$
$$100 \equiv 1 \ (\text{mod } 3).$$

The fraction $\dfrac{1}{9} = 0.111 \cdots$ also is constant that is, it repeats in blocks of unit length as does the sequence of remainders:

$$1 \equiv 1 \ (\text{mod } 9);$$
$$10 \equiv 1 \ (\text{mod } 9);$$
$$100 \equiv 1 \ (\text{mod } 9).$$

The situation is much the same with $\dfrac{1}{6} = 0.166\ 666 \cdots$ in which the decimal expansion is constant, after an initial, singular digit is passed. The same pattern is found among the congruences

$$1 \equiv 1 (\text{mod } 6); \qquad\qquad 10 \equiv 4 (\text{mod } 6);$$
$$10^2 \equiv 4 (\text{mod } 6); \qquad\qquad 10^3 \equiv 4 (\text{mod } 6);$$
$$\text{etc.}$$

To consider an example in which the repeating blocks are slightly longer, take the congruences of 37:

$$1 \equiv 1 \pmod{37};$$
$$10 \equiv 10 \pmod{37};$$
$$100 \equiv 26 \pmod{37};$$
$$1000 \equiv 1 \pmod{37};$$

all of them follow the same pattern as appears in the decimal development

$$\frac{1}{37} = 0.027\ 027\cdots.$$

In summary, then, we have

period of length 1	$\dfrac{1}{3}, \dfrac{1}{9}$
period of length 2	$\dfrac{1}{11}$
period of length 3	$\dfrac{1}{37}$
period of length 6	$\dfrac{1}{7}, \dfrac{1}{13}$

28. Repeating Digits and Remainders Mod 3

Is it possible to obtain decimals that repeat in blocks of any given length? Any power of 10 satisfies $10^n \equiv 1 \pmod{3}$. We can obtain repeating decimals of any length by taking reciprocals of numbers of the form $10^n - 1$.

period	number	factors
1	10^1-1	3^2
2	10^2-1	$3^2 \cdot 11$
3	10^3-1	$3^3 \cdot 37$
4	10^4-1	$3^2 \cdot 11 \cdot 101$
5	10^5-1	$3^2 \cdot 41 \cdot 271$
6	10^6-1	$3^3 \cdot 7 \cdot 11 \cdot 13 \cdot 37$
7	10^7-1	$3^2 \cdot 239 \cdot 4649$
8	10^8-1	$3^2 \cdot 11 \cdot 73 \cdot 101 \cdot 137$
9	10^9-1	$3^4 \cdot 37 \cdot 333,667$

For example $10^3 - 1 = 999$, will have a reciprocal, $\dfrac{1}{999}$, whose decimal expansion repeats in blocks of three.

29. A Survey of Decimal Repetition

In this section a survey is provided of the behavior of the decimal expansions of the smallest positive integers.

$\frac{1}{2}$	one digit
$\frac{1}{3}$	a period of length one
$\frac{1}{4}$	two digits
$\frac{1}{5}$	one digit
$\frac{1}{6}$	one digit, then a period of one
$\frac{1}{7}$	a period of six
$\frac{1}{8}$	three digits
$\frac{1}{9}$	a period of one
$\frac{1}{10}$	one digit
$\frac{1}{11}$	a period of two
$\frac{1}{12}$	two digits, then a period of one
$\frac{1}{13}$	a period of six
$\frac{1}{14}$	one digit, then a period of six
$\frac{1}{15}$	one digit, then a period of one
$\frac{1}{16}$	four digits
$\frac{1}{17}$	a period of sixteen
$\frac{1}{18}$	one digit, then a period of one
$\frac{1}{19}$	eighteen digits
$\frac{1}{20}$	two digits

A Peculiar Mental Calculation

30. A Remark Concerning Fifth Roots

It is a fact that the sum of the digits of 205,962,976 is 46. Furthermore,

$$\sqrt[5]{205,962,976} = 46.$$

Certainly the remainders mod 9 of 46 and 205,962,976 are the same; both equal 1.

To take another example of a fifth root, let us consider

$$\sqrt[5]{1,419,857} = 17.$$

Here the sum of the digits of 1,419,857 is 35 ≡ 8 (mod 9). So again 1,419,857 and 17 have the same remainder mod 9.

There are some general patterns that arise with fifth roots.

First Statement: The last digit of the original number will be the last digit of the fifth root.

The justification for this is to be found in the following congruences:

$n \equiv 2 \pmod{10}$	$n \equiv 3 \pmod{10}$	$n \equiv 4 \pmod{10}$
$n^2 \equiv 4 \pmod{10}$	$n^2 \equiv 9 \pmod{10}$	$n^2 \equiv 6 \pmod{10}$
$n^3 \equiv 8 \pmod{10}$	$n^3 \equiv 7 \pmod{10}$	$n^3 \equiv 4 \pmod{10}$
$n^4 \equiv 6 \pmod{10}$	$n^4 \equiv 1 \pmod{10}$	$n^4 \equiv 6 \pmod{10}$
$n^5 \equiv 2 \pmod{10}$	$n^5 \equiv 3 \pmod{10}$	$n^5 \equiv 4 \pmod{10}$

$n \equiv 5 \pmod{10}$	$n \equiv 7 \pmod{10}$	$n \equiv 9 \pmod{10}$
$n^2 \equiv 5 \pmod{10}$	$n^2 \equiv 9 \pmod{10}$	$n^2 \equiv 1 \pmod{10}$
$n^3 \equiv 5 \pmod{10}$	$n^3 \equiv 3 \pmod{10}$	$n^3 \equiv 9 \pmod{10}$
$n^4 \equiv 5 \pmod{10}$	$n^4 \equiv 1 \pmod{10}$	$n^4 \equiv 1 \pmod{10}$
$n^5 \equiv 5 \pmod{10}$	$n^5 \equiv 7 \pmod{10}$	$n^5 \equiv 9 \pmod{10}$

Second Statement: The sum of the digits of a fifth root can be determined mod 9 from the sum of the digits of its fifth power.

$n \equiv 7 \pmod{10}$	$n \equiv 8 \pmod{10}$	$n \equiv 9 \pmod{10}$
$n^2 \equiv 9 \pmod{10}$	$n^2 \equiv 4 \pmod{10}$	$n^2 \equiv 1 \pmod{10}$
$n^3 \equiv 3 \pmod{10}$	$n^3 \equiv 2 \pmod{10}$	$n^3 \equiv 9 \pmod{10}$
$n^4 \equiv 1 \pmod{10}$	$n^4 \equiv 6 \pmod{10}$	$n^4 \equiv 1 \pmod{10}$
$n^5 \equiv 7 \pmod{10}$	$n^5 \equiv 8 \pmod{10}$	$n^5 \equiv 9 \pmod{10}$

Second Statement: The sum of the digits of a fifth root can be determined mod 9 from the sum of the digits of its fifth power.

This relationship is shown by the calculations mod 9 given below. The sum of the digits is s.

$s \equiv 1 \pmod 9$	$s \equiv 2 \pmod 9$	$s \equiv 3 \pmod 9$
$s^2 \equiv 1 \pmod 9$	$s^2 \equiv 4 \pmod 9$	$s^2 \equiv 0 \pmod 9$
$s^3 \equiv 1 \pmod 9$	$s^3 \equiv 8 \pmod 9$	$s^3 \equiv 0 \pmod 9$
$s^4 \equiv 1 \pmod 9$	$s^4 \equiv 7 \pmod 9$	$s^4 \equiv 0 \pmod 9$
$s^5 \equiv 1 \pmod 9$	$s^5 \equiv 5 \pmod 9$	$s^5 \equiv 0 \pmod 9$

$s \equiv 4 \pmod 9$	$s \equiv 5 \pmod 9$	$s \equiv 6 \pmod 9$
$s^2 \equiv 7 \pmod 9$	$s^2 \equiv 7 \pmod 9$	$s^2 \equiv 0 \pmod 9$
$s^3 \equiv 1 \pmod 9$	$s^3 \equiv 8 \pmod 9$	$s^3 \equiv 0 \pmod 9$
$s^4 \equiv 4 \pmod 9$	$s^4 \equiv 4 \pmod 9$	$s^4 \equiv 0 \pmod 9$
$s^5 \equiv 7 \pmod 9$	$s^5 \equiv 2 \pmod 9$	$s^5 \equiv 0 \pmod 9$

$s \equiv 7 \pmod 9$	$s \equiv 8 \pmod 9$	$s \equiv 0 \pmod 9$
$s^2 \equiv 4 \pmod 9$	$s^2 \equiv 1 \pmod 9$	$s^2 \equiv 0 \pmod 9$
$s^3 \equiv 1 \pmod 9$	$s^3 \equiv 8 \pmod 9$	$s^3 \equiv 0 \pmod 9$
$s^4 \equiv 7 \pmod 9$	$s^4 \equiv 1 \pmod 9$	$s^4 \equiv 0 \pmod 9$
$s^5 \equiv 4 \pmod 9$	$s^5 \equiv 8 \pmod 9$	$s^5 \equiv 0 \pmod 9$

This information is summarized in the table that follows on the next page. The double entries in the left-hand column arise because the sum of the digits in a number is not exactly the same thing as its remainder mod 9. When the sum of the digits is calculated, you may obtain a 9 instead of the actual remainder of 0.

a^5	a
1	1
5	2
0, 9	3
7	4
2	5
0, 9	6
4	7
8	8

The facts in this table allow us to find fifth roots quickly when the root has only two digits. For example: Find $\sqrt[5]{6436343}$. The sum s has $s^5 = 2$ (mod 9); from the table $s = 5$. So the root is 23. We know this as soon as we recognize that the root must be in the 20's, and that the digits must have a sum of 5.

Problem: Find $\sqrt[5]{5153632}$.

Solution: For this sum, $s^5 = 7$ (mod 9). So from the preceeding table $s = 4$. The root is 22.

Fifth roots of numbers greater than one hundred can also be found with equal ease, provided one is able to obtain the first digit by knowing the intervals in which fifth powers lie.

V. The Binary System

Binary Numbers

31. The Binary System is Based on Powers of Two

$$2^0 = 1$$
$$2^1 = 2 \qquad\qquad 2^{-1} = 0.5$$
$$2^2 = 4 \qquad\qquad 2^{-2} = 0.25$$
$$2^3 = 8 \qquad\qquad 2^{-3} = 0.125$$
$$2^4 = 16 \qquad\qquad 2^{-4} = 0.0625$$
$$2^5 = 32 \qquad\qquad 2^{-5} = 0.03125$$
$$2^6 = 64 \qquad\qquad 2^{-6} = 0.0015625$$
$$2^7 = 128 \qquad\qquad 2^{-7} = 0.00078125$$
$$2^8 = 256 \qquad\qquad 2^{-8} = 0.000390625$$
$$2^9 = 512 \qquad\qquad 2^{-9} = 0.0001953125$$
$$2^{10} = 1024 \qquad\qquad 2^{-10} = 0.00009765625$$

32. Decimal to Binary

Problem: Express in binary form: 846.

Solution: Subtract the largest power of two from 846 that will yield a positive difference. Then repeat the process on the differences.

$$
\begin{array}{rr}
 & 846 \\
2^9 = & \underline{512} \\
 & 334 \\
2^8 = & \underline{256} \\
 & 78 \\
2^6 = & \underline{64} \\
 & 14 \\
2^3 = & \underline{8} \\
 & 6 \\
2^2 = & \underline{4} \\
 & 2 \\
2^1 = & \underline{2} \\
 & 0
\end{array}
$$

Finally, we need to record the powers of two that have been found in 846. Since 846 contained 512, the ninth power of 2, a

1 is recorded as the 10th binary digit of 846 (the first digit represents 2 to the power zero, so the count is "off by one" and the 9th power of two will be marked in the tenth position reading right to left).

9	8	7	6	5	4	3	2	1	0	........powers of two
1	1	0	1	0	0	1	1	1	0	........bits

Since the word "digit" suggests that there are ten objects, we have adopted the usage of computer programmers and refer to a binary digit as a "bit".

So 846 is 1101001110 when expressed in binary notation. This completes the solution of the problem.

Binary notation has the advantage of expressing numbers using only the two symbols "0" and "1"; it has the disadvantage of requiring long expressions.

Problem: As a second example, express 367 in binary form.

$$\begin{array}{r} 367 \\ \underline{256} \\ 111 \\ \underline{64} \\ 47 \\ 32 \\ 15 \\ \underline{8} \\ 7 \\ \underline{4} \\ 3 \\ \underline{2} \\ 1 \end{array}$$

8	7	6	5	4	3	2	1	0	powers of two
1	0	1	1	0	1	1	1	1	bits

33. Binary to Decimal

The reverse process is carried out in a similar way.

Problem: Express 100111010 in the usual base ten notation.

Solution: First assign binary exponents to these binary digits or "bits."

$$\frac{8\ \ 7\ \ 6\ \ \ 5\ \ 4\ \ 3\ \ 2\ \ \ 1\ \ 0}{1\ \ 0\ \ 0\ \ \ 1\ \ 1\ \ 1\ \ 0\ \ \ 1\ \ 0}$$

Next, insert suitable power of two wherever the bit 1 appears.

$$2^8 = 256$$
$$2^5 = 32$$
$$2^4 = 16$$
$$2^3 = 8$$
$$2^1 = \underline{2}$$
$$314$$

Problem: Translate 1100111011 into the ordinary decimal, base ten, notation.

Solution:

$$\frac{9\ \ 8\ \ 7\ \ 6\ \ 5\ \ 4\ \ 3\ \ 2\ \ 1\ \ 0}{1\ \ 1\ \ 0\ \ 0\ \ 1\ \ 1\ \ 1\ \ 0\ \ 1\ \ 1}$$

$$2^9 + 2^8 + 2^5 + 2^4 + 2^3 + 2^1 + 2^0 =$$

$$512 + 256 + 32 + 16 + 8 + 2 + 1 = 827.$$

This completes the problem.

To provide another example, the binary number 1111111111 is 1023, which is one less than 2^{10}.

34. Binary Fractions

Sample Problem: Translate 48.40625 into binary notation.

Solution:

$$
\begin{array}{rll}
 & 48\ .\ 40625 & \\
2^5 = & \underline{32\ .\ 25} & = 2^{-2} \\
 & 16\ .\ 15625 & \\
2^4 = & \underline{16\ .\ 125} & = 2^{-3} \\
 & .\ 03125 & \\
 & \underline{.\ 03125} & = 2^{-5} \\
 & .\ 0 &
\end{array}
$$

The fractional powers of two are written at the right; 32.25 = $2^5 + 2^{-2}$. From this calculation we obtain the result:

$$\underline{5\ 4\ 3\ 2\ 1\ 0\ -1\ -2\ -3\ -4\ -5}$$
$$1\ 1\ 0\ 0\ 0\ 0\ 0\ 1\ 1\ 0\ 1$$

So 48.40625 is 110000.01101 in the binary system.

35. Binary Multiplication

To multiply $12 \times 9 = 108$ in base two, proceed as in base ten

$$1100 \times 1001$$
$$1100$$
$$1100$$
$$\rule{3cm}{0.4pt}$$
$$1101100$$

[Here, as in other places in Gisela O'Neil's notes the multiplications are arranged differently from the way it is done. We would ordinarily write it as shown below. I do not know whether she was hasty in writing or whether von Baravalle had an Austrian style of writing his products. The fact remains, however the computation is written, that the very same methods used in decimal arithmetic will work for binary arithmetic too. – Ed.]

In the usual notation it would appear as $\quad$ 1100
$$\underline{1001}$$
$$1100$$
$$\underline{1100\quad .}$$
$$1101100$$

The result may be checked as before.

Label: $\underline{6\ 5\ 4\ 3\ 2\ 1\ 0}$
$$1\ 1\ 0\ 1\ 1\ 0\ 0$$

Then add the corresponding powers of two:

$$2^6 + 2^5 + 2^3 + 2^2 = 64 + 32 + 8 + 4 = 108.$$

We can do the reverse equally well: begin with a decimal calculation, then translate the problem into binary notation. Let us choose a product, 112×72, for example,

112×72 In binary notation $112 = 64 + 32 + 16 = 111000$.
784   While $72 = 64 + 8$ or 1001000
224
8064

[Here again we have preserved the unusual form of the calculations that appear in the notes. – Ed.]

In binary form the calculation appears as follows:

$$\underline{1110000 \times 1001000}$$
$$1110000$$
$$1110000$$

Adding these columns gives the product

$$1111110000000.$$

Now, this answer can be returned to decimal form and compared with the result obtained by doing the calculation in ordinary decimals.

12 11 10 9 8 7 6 5 4 3 2 1 0
1 1 1 1 1 1 0 0 0 0 0 0 0

$2^{12} + 2^{11} + 2^{10} + 2^9 + 2^8 + 2^7 =$

$= 4096 + 2048 + 1024 + 512 + 256 + 128 = 8064.$

36. Binary Addition

Adding $103 + 17 + 36 = 156$ in binary form resembles an ordinary addition.

Breaking the summands down into powers of two gives:
$103 = 64 + 32 + 4 + 2 + 1;$ $17 = 16 + 1;$
$36 = 32 + 4$ and $156 = 128 + 16 + 8 + 4.$

These give us the binary numbers to be added.

```
    1100111
      10001
     100100
  10011100
```

Similarly
```
    1100001
     101001
   10000001
      10001
  100011100
```

corresponds to 97 + 41 + 129 + 17 = 284.

37. Binary Subtraction

There are two methods of subtracting in common use: the method of borrowing, in which one is transferred, and the Austrian method , in which one is added.

```
 ¹2̶4        24
  15         1̍5
```

These methods serve for binary subtraction too. Whichever of these methods of subtraction is chosen, it can be seen to have a binary analog.

example:		example:	
110101	53	102	1100110
− 11011	− 27	− 47	− 101111
11010	26	55	110111

The inner subtractions are ordinary decimal subtractions. The outer pair are the binary versions of these two examples.

38. Binary Division

$$27 : 9 = 3 \qquad\qquad 91 : 7 = 13$$

$11011 : 1001 = 11$	$1011011 : 111 = 1101$
1001	111
1001	1000
	111
	111

[Here as in other places von Baravalle arranges his working computations in an European style. These particular divisions are carried out exactly as Americans are accustomed to do them, except that the divisors and quotients are written down in a different place on the page. – Ed.]

39. Patterns in Binary Numbers

Compare the binary numbers with their base ten counterparts.

1	1	9	1001
2	10	10	1010
3	11	11	1011
4	100	12	1100
5	101	13	1101
6	110	14	1110
7	111	15	1111
8	1000		

Using the list, let us record where 1's appear in the binary expressions of the various numbers.

right digit	second digit	third digit	left digit
1	2	4	8
3	3	5	9
5	6	6	10
7	7	7	11
9	10	12	12
11	11	13	13
13	14	14	14
15	15	15	15

Interesting observations can be made from a list of this kind. Odd numbers, of course, are exactly those having a one as their right hand numeral. A number which is one less than a power of two, like 15, will have ones in every position.

Now, let us take an arbitrary number, 87 will do, and express it in binary form: 1010111.

Dividing it in half successively gives

$$87 = 43 \cdot 2 + 1$$
$$43 = 21 \cdot 2 + 1$$
$$21 = 10 \cdot 2 + 1$$
$$10 = 5 \cdot 2 + 0$$
$$5 = 2 \cdot 2 + 1$$
$$2 = 1 \cdot 2 + 0$$
$$1 = 0 \cdot 2 + 1$$

Reading this list of remainders from bottom to top gives the binary digits of 87 in order, taken from right to left. This fact is used in the method of Russian multiplication.

Russian Multiplication

40. How to Multiply by Doubling

Doubling a number was once considered a separate operation, distinct from multiplication. Russian multiplication is a method of multiplication that is more laborious than the usual ones, but it requires less memorizing. Multiplication tables are not needed. Only doubling is used. The method is based on binary arithmetic. As an example take the product

$$87 \cdot 59 = 5133.$$

Form the sequence of halves that is given at the end of the previous section; this gives the numbers 43, 21, 10, 5, 2 and 1. As 87 is successively divided in half, 59 is doubled.

At certain positions in the list numbers appear that correspond to zeros in the binary expansion 1010111 of 87.

These are also positions that have remainders of 0 in the breakdown of 87. These positions are crossed out of the list.

87	·	59
43		118
21		236
10		~~472~~
5		944
2		~~1888~~
1		3776
		5133

In the left-hand column 87 has been sucessively cut in half. When the number is odd, it is decreased by one and then cut in half. In certain positions it is even; it can be halved directly. At those places a number in the right-hand column is crossed out. The right-hand column itself is obtained by doubling the other factor, 59.

Finally, the right-hand column is added but the crossed out numbers are omitted from the sum. This gives 5133, the product of 87 and 59.

The method can be checked by applying it to the product of the same factors taken in the reverse order, 59·87.

$$59 = 111011 \text{ (base 2)}$$

59	87	$59 = 29 \cdot 2 + 1$
29	174	$29 = 14 \cdot 2 + 1$
14	~~348~~	$14 = 7 \cdot 2 + 0$
7	696	$7 = 3 \cdot 2 + 1$
3	1392	$3 = 1 \cdot 2 + 1$
1	2784	$1 = 0 \cdot 2 + 1$
	5133	

41. The Ternary System

The ternary system is based on the powers of three and makes use only of the digits 0, 1, 2.

$3^0 =$	1	$3^6 =$	729
$3^1 =$	3	$3^7 =$	2,187
$3^2 =$	9	$3^8 =$	6,561
$3^3 =$	27	$3^9 =$	19,683
$3^4 =$	81	$3^{10} =$	59,049
$3^5 =$	243	$3^{11} =$	177,147

For example, $107 = 10222_{(3)}$ and $59 = 2012_{(3)}$.

As in the case of binary numbers we may arrive at these expressions by calculating remainders mod 3:

$$107 = 35 \cdot 3 + 2 \qquad 59 = 19 \cdot 3 + 2$$
$$35 = 11 \cdot 3 + 2 \qquad 19 = 6 \cdot 3 + 1$$
$$11 = 3 \cdot 3 + 2 \qquad 6 = 2 \cdot 3 + 0$$
$$3 = 1 \cdot 3 + 0 \qquad 2 = 0 \cdot 3 + 2$$
$$1 = 0 \cdot 3 + 1$$

The process of Russian multiplication can also be carried out using the ternary expansions.

VI. Euclid's Algorithm

Greatest Common Divisors

42. The GCD

Euclid studied pure numbers; these studies are older than any algebraic methods. Euclid's algorithm is a calculation that repeats an arithmetic procedure. It can be described in purely arithmetical terms without algebraic notation.

First, consider a typical fraction:

$$\frac{76,084}{63,020}$$

Factor these two numbers into primes obtaining:

$$76,084 = 2^2 \cdot 23 \cdot 827, \text{ and } 63,020 = 2^2 \cdot 5 \cdot 23 \cdot 137.$$

The factors that appear in common can be canceled from the fractions leaving:

$$\frac{827}{137 \cdot 5}$$

The largest number that could be canceled was $2^2 \cdot 23 = 92$; this number is said to be the *greatest common divisor* of 76,084 and 63,020.

There is a computational algorithm that leads to this greatest common divisor, 92. First, observe that 63,020 goes into 76,048 but once. We may write this in three staggered columns.

$$76,084$$
$$63,020 \qquad 1$$
$$13,064$$

The difference, 13,064, is recorded beneath the original entry. We divided the middle column into the left one and recorded the remainder. Now, we do the reverse and divide the leftmost column into the middle, obtaining:

$$
\begin{array}{lll}
76{,}084 & & \\
& 63{,}020 & 1 \\
13{,}064 & & 4 \\
& 10{,}764 & 1 \\
\end{array}
$$

since the quotient of 63,020 by 13064 is 4 and the remainder is 10764.

Continuing the algorithmic process, we take 10,764 into 13,064 and record the quotient and remainder.

$$
\begin{array}{lll}
76{,}084 & & \\
& 63{,}020 & 1 \\
13{,}064 & & 4 \\
& 10{,}764 & 1 \\
2{,}300 & & 4 \\
\end{array}
$$

Yet another repetition gives:

$$
\begin{array}{lll}
76{,}084 & & \\
& 63{,}020 & 1 \\
13{,}064 & & 4 \\
& 10{,}764 & 1 \\
2{,}300 & & 4 \\
& 1{,}564 & 1 \\
\end{array}
$$

The numbers are decreasing as we make up these columns and so the process must end sometime. Continuing, we soon arrive at

$$
\begin{array}{lll}
76{,}084 & & \\
& 63{,}020 & 1 \\
13{,}064 & & 4 \\
& 10{,}764 & 1 \\
2{,}300 & & 4 \\
& 1{,}564 & 1 \\
736 & & 2 \\
& \underline{92} & 8 \\
\end{array}
$$

Upon dividing 92 into 736 we find that the quotient is 8 and the remainder is 0. So the process halts here. There is no way to continue by dividing 0 into 92.

This algorithmic process has led us to the greatest common divisor of 76,084 and 63,020, namely 92.

43. The Algorithm at Work

As a second example let us simplify the fraction $\dfrac{2754}{1632}$.
By listing the prime factors of these two numbers we may simplify the fraction without using Euclid's algorithm.

2754	2		1632	2
1377	3		816	2
459	3		408	2
153	3		204	2
51	3		102	2
17	17		51	3
1			17	17

After the like factors have been canceled as shown, the result can be expressed as:

$$\frac{2754}{1632} = \frac{3^3}{2^4} = \frac{27}{16}.$$

Plainly, the greatest common divisor of these two numbers , 2754 and 1632, is found in the common factors that have been canceled: $2 \cdot 3 \cdot 17 = 102$. Again the algorithmic process of successive divisions also leads to this same result:

```
2754
                1632        1
        1122                1
                 510        2
         102                5
```

The process stops at 102 because 102 divides 510 at the previous step. As soon as there is no renainder, the algorithm halts.

To choose another example, form the fraction

$$\frac{3552}{7400} = \frac{2^5 \cdot 3 \cdot 37}{2^3 \cdot 5^2 \cdot 37} = \frac{2^2 \cdot 3}{5^2} = \frac{12}{25}.$$

In simplifying this fraction $2^3 \cdot 37 = 296$ was canceled. So also 296 is the *gcd*, the greatest common divisor, of 3552 and 7400. In this instance the algorithm in which we carry out successive divisions is quite brief. It halts when one of the divisions can be done without a remainder.

$$7400$$
$$\qquad\qquad 3552 \qquad 2$$
$$\underline{296} \qquad\qquad 12$$

Again the algorithm gives the gcd of the two numbers. It halts because 296 divides evenly into 3552.

44. Proving the Algorithm

The term *algorithm* refers to the successive repetition of the same procedure. Now, return to the first example of this algorithm, the calculation of the gcd of 76,084 and 63,020 that appears in section 42. Adding some detail to the calculation we obtain the sequence

$$76,084 = 63,020 \cdot 1 + 13,064$$

$$63,020 = 13,064 \cdot 4 + 10,764$$

$$13,064 = 10,764 \cdot 1 + 2,300$$

$$10,764 = 2,300 \cdot 4 + 1,564$$

$$2,300 = 1564 \cdot 1 + 736$$

$$1,564 = 736 \cdot 2 + 92$$

$$736 = \underline{92} \cdot 8$$

Looking at these equations we see that the factors that are in common to the original entries, 76084 and 63020, must run through the computation and appear as factors of the three terms in each of the lines. So 92 must be a factor common to both.

Least Common Multiple

45. The LCM

Begin with the problem of adding fractions:

$$\frac{1}{24} + \frac{1}{30} + \frac{1}{48}$$

These numbers factor into primes giving: $24 = 2^3 \cdot 3$;
$30 = 2 \cdot 3 \cdot 5$;
$48 = 2^4 \cdot 3$.

We could take some common factors out simultaneously.

			factor
24	30	48	2
12	15	24	2
6	15	12	2
3	15	6	3
1	5	2	

The factor 2 was recorded on the right when it was removed from all three of the numbers 24, 30, and 48 simultaneously, and similarly for the other entries. The second 2 in the right hand column was removed from 12 and 24.

From this table we find that the *common denominator* of these numbers is

$$2 \cdot 2 \cdot 2 \cdot 2 \cdot 3 \cdot 5 = 16 \cdot 3 \cdot 5.$$

This is the *least common multiple* of 24, 30, and 48, since it includes the prime factors of each.

The table that follows shows how one would find the least common multiple of all the numbers from 1 to 10, using this same scheme of removing factors from several numbers at each step.

1	2	3	4	5	6	7	8	9	10	factor
										2
	1	3	2	5	3	7	4	9	5	2
		3	1	5	3	7	2	9	5	3
		1		5	1	7	2	3	5	5
				1		7	2	3	1	

So the least common multiple, the *lcm*, of these first ten numbers is

$$lcm = 2 \cdot 2 \cdot 2 \cdot 3 \cdot 3 \cdot 5 \cdot 7.$$

This number would also be the common denominator if we were to add the reciprocals of the first ten integers.

VII. Indeterminate Problems

Diophantine Problems of Euler

46. A Problem Demanding Integral Solutions

This problem was posed by the Swiss mathematician Leonhard Euler.

Problem: Find all the possible solutions: divide 25 into two numbers, one of which is divisible by two and the other by three.

The possible solutions could be listed and then checked to see which ones involved multiples of 2 and 3.

25 = 1 + 24 ✖		25 = 7 + 18 ✖
25 = 2 + 23 ✖		25 = 8 + 17 ✖
25 = 3 + 22 ✓		25 = 9 + 16 ✓
25 = 4 + 21 ✓		25 = 10 + 15 ✓
25 = 5 + 20 ✖		25 = 11 + 14 ✖
25 = 6 + 19 ✖		25 = 12 + 13 ✖

This is not a satisfactory method, however, because it is not systematic.

Solution: First, the problem must be expressed as an equation. There will be two unknowns.

$$2x + 3y = 25$$

Or, equivalently: $\quad 2x = -3y + 25.$

Now, we divide by 2, but we also separate the y's.

$$x = -y + 12 + \frac{1-y}{2}.$$

To meet the conditions of the problem, $\dfrac{1-y}{2} = t$ cannot be a fraction. Solving this latter equation for y gives:

$$y = 1 - 2t.$$

This means that $x = -1 + 2t + 12 + t$. And we obtain the pair of conditions:

$$x = 3t + 11$$

$$y = 1 - 2t$$

Now, if the solution y is to be positive, then $1 - 2t > 0$ or $t < \frac{1}{2}$. And, if x is to be positive as well, then $3t > -11$ or, equivalently, $t > -3\frac{2}{3}$. These conditions restrict our possible choices of t to

$$t = 0, \; -1, \; -2, \; -3$$

$t = 0$	$x = 11$	$y = 1$	$22 + 3 = 25$
$t = -1$	$x = 8$	$y = 3$	$16 + 9 = 25$
$t = -2$	$x = 5$	$y = 5$	$10 + 15 = 25$
$t = -3$	$x = 2$	$y = 7$	$4 + 21 = 25$

No further solutions are possible.

47. Another Problem of Euler

Problem: A man bought horses and cows for $1770, more cows than horses. One horse cost $31 and one cow cost $21. How many horses and how many cows did he buy?

Solution: The problem gives the equation

$$31x + 21y = 1770.$$

In addition, it is clear from the meaning of the problem itself that the following two conditions should also be placed on the solutions:

1) no fractions,

2) no negative numbers.

Solving the equation for y gives

(✿) $$y = -x + 84 + \frac{6 - 10x}{21}.$$

As before, substitute a single symbol for the fraction that appears in this expression. By the conditions of the problem, all three terms in the expression for y must be integers. Thus, we my assume that there is an integer t such that

$$6 - 10x = 21t.$$

Solving this expression for x gives

$$x = -2t + \frac{6 - t}{10}.$$

Continuing the same line of thought, we see that the two terms on the right hand side must be integers since x itself is, so we may assume that there is an integer u with

$$\frac{6 - t}{10} = u.$$

This time, solving for t, there is an expression, $t = 6 - 10u$, in which fractions no longer appear. This allows us to return to equation (✿) above and obtain expressions for y and x:

$$y = -(-2t + u) + 84 + t = 3t - u + 84 =$$

$$= 3(6 - 10u) - u + 84 = 102 - 31u;$$

$$x = -2t + u = -2(6 - 10u) + u = 21u - 12.$$

Since y is positive, $102 - 31u > 0$. This means that u can be 3 at most. The condition that x be positive too requires that $21u - 12 > 0$, which forces u to be at least 1. Thus 1, 2, and 3 are the only possible values for u.

$21u - 12 > 0$, which forces u to be at least 1. Thus 1, 2, and 3 are the only possible values for u.

u	x	y
1	9	71
2	30	40
3	51	9

The last row of the table, in which $u = 3$, gives the only solution which meets the condition that there be more horses than there are cows.

Diophantine Problems from India

48. An Indian Problem

Problem: Find the number which will be exhausted when it is multiplied by 221, 65 is added, and the sum is divided by 195.

The word "exhausted " here means evenly divisible by 195, so that the result is an integer.

Solution: We divide the equation

$$221x + 65 = 195y$$

by 13 to obtain:

$$17x + 5 = 15y.$$

Proceeding by the method of making fractions disappear, we solve to obtain

$$y = x + \frac{2x + 5}{15}$$

and then introduce $15t = 2x + 5$. This gives

$$x = 7t - 2 + \frac{t - 1}{2}.$$

Since there are still fractions, the process must continue. Let $t - 1 = 2u$. This time when we solve for t we obtain an expression $t = 2u + 1$ lacking fractions. So the original variables x and y may be replaced by u in an algebraic expression without any fractional part.

$$x = 7(2u + 1) - 2 + u = 15u + 5;$$

$$y = 15u + 5 + 2u + 1 = 17u + 6.$$

Next, use the facts that x and y are positive to write

$$15u + 5 > 0 \text{ and } 17u + 6 > 0.$$

Together these inequalities require that $u > -1/3$. This means that there are an infinite number of solutions $u = 0, 1, 2, 3,$ and so on. Take the smallest possible solution, $u = 0$; then $x = 5$ and $y = 6$. A quick calculation shows that the conditions of the problem are met.

49. Another Example

Problem: Divide 100 into two parts one of which is divisible by 7 and the other by 11.

Solution: As before introduce an equation $7x + 11y = 100$. Solving for x gives

$$x = 14 - y + \frac{2 - 4y}{7}$$

and to eliminate fractions put $7t = 2 - 4y$. This gives

$$y = -2t + \frac{2 + t}{4}$$

which still contains a fractional expression; so set $4u = 2 + t$. At last $t = -2 + 4u$ is free of fractions. Unwinding the substitutions gives:

$$x = 8 + 11u \; ; \; y = 4 - 7u.$$

This means that $4/7 > u > -8/11$, so that u must equal 0. So, we have that $x = 8$ and $y = 4$. The problem asks for the partition $100 = 56 + 44$.

50. In an Inn

Diophantine problems, those requiring whole number solutions, were particularly common in India. There was a collection of them made by Claude Caspar Bachet, Sieur de Meritiac, 1587-1638. The following problem appeared in his collection:

Problem: A party of 41 persons, men, women, and children, take part in a meal in an inn. Their bill comes to 40 units of money. Each man pays 4, each woman pays 3, and each child 1/3. How many men, women, and children took part?

Solution: Let us begin with the equations

$$x + y + z = 41;$$

$$4x + 3y + \frac{z}{3} = 40.$$

It is best to eliminate fractions from this second equation and express it as

$$12x + 9y + z = 120.$$

By subtracting the first equation from this one, the z disappears, leaving

$$11x + 8y = 79.$$

Solving for y and then substituting for fractional terms gives

$$y = -x + 10 + \frac{-3x - 1}{8} = -x + 10 + t$$

where $-3x - 1 = 8t$.
Which is to say that

$$x = -3t + \frac{t-1}{3}.$$

Now we continue the same process and eliminate the fraction. To do this let $3u = t - 1$. When we solve for t, we obtain an expression, $t = 3u + 1$, that is free of fractional terms. We may now work back to express x and y in terms of u. This gives:

$$x = -3t + u = -8u - 3;$$

$$y = -x + 10 + t = -(-3t + u) + 10 + (3u + 1) =$$

$$= 3(3u + 1) + 2u + 11 = 11u + 14.$$

Since x and y are positive integers, that is $-8u - 3 > 0$ and $11u + 14 > 0$, we obtain the conditions that $-3/8 > u$ and $u > -14/11$. This means that u must be -1. This in turn gives $x = 5$ and $y = 3$; so the problem asks for the sums:

$$5 + 3 + 33 = 41$$
$$\text{and} \quad 20 + 9 + 11 = 40.$$

The first of these sums gives the composition of the party – men, women, and children respectively: The second sum gives the distribution of costs for the meal.

51. The Indian Style

Problems of the type that we are considering were given in a beautiful style in the Orient. This can be seen in a striking way in a problem of Mahaviracarya.

Into the bright and refreshing outskirts of a forest which were full of numerous trees with their branches bent down with the weight of flowers and fruits, trees such as jambu trees, date-palms, hintala trees, palmyras, punnaga trees and mango trees – filled with the many sounds of crowds of parrots and cuckoos found near springs containing lotuses with bees roaming around them – a number of travelers entered with joy. There were 63 equal heaps of plantain fruits put together and seven single fruits. These were divided evenly among 23 travelers. Tell me now the number of fruits in each heap.

Solution: The problem asks for integral solutions of

$$63x + 7 = 23y.$$

Since $y = 3x + \dfrac{7 - 6x}{23}$, we substitute $7 - 6x = 23t$.

Solving for x gives

$$x = -4t + 1 + \frac{t+1}{6}.$$

Repeating the process, put $t + 1 = 6u$. Since $t = 6u - 1$, one may work back to obtain: $x = -23u + 5$ and $y = -63u + 14$. Because x and y are positive, it must be that $-23u + 5 > 0$ and $-63u + 14 > 0$, so that $5/23 > u$ and $14/63 > u$ respectively. This means that $u = 0, -1, -2, -3, \cdots$. The smallest value of x and y will occur when $u = 0$; This gives $x = 5$ and $y = 14$.

This solution can be checked:

$$(63 \times 5) + 7 = 315 + 7 = 322 = 23 \times 14.$$

There is an abbreviated approach to this problem using congruences. Initially there was the equation

$$63x + 7 = 23y.$$

Since 63 and 7 are both divisible by 7, and 23 is prime, it must be that y is divisible by 7. So there must be an integer y' with $y = 7y'$. The problem may be expressed as

$$9x + 1 = 23y'.$$

This latter equation may be treated by the methods of this chapter.

Monetary Units

52. Making Change

Problem: In how many ways can twenty-five cents be changed into nickels and dimes?

Intuitively we have the solutions

$$5 + 5 + 5 + 5 + 5$$
$$5 + 5 + 5 + 10$$
$$5 + 10 + 10$$

Solution: Express the problem in the equation

$$5x + 10y = 25.$$

Dividing by five gives $\qquad x + 2y = 5;$

or $\qquad x = -2y + 5.$

Since x must be positive $\qquad -2y + 5 > 0$ or $y < 2.5$.
    This requires that $y = 0$, 1, or 2. So x is 5, 3, or 1.

53. What if Pennies are Allowed?

If we allow that pennies are admitted into the ways of changing a quarter, then there are twelve possible ways to make change.

dimes	nickels	pennies
2	1	0
2	0	5
1	3	0
1	2	5
1	1	10
1	0	15
0	5	0
0	4	5
0	3	10
0	2	15
0	1	20
0	0	25

When we approach the problem systematically, we begin with an equation in three unknowns.

$$x + 5y + 10z = 25.$$

$$x = -5y - 10z + 25.$$

Since x is not negative, it must be that

$$25 > 5y + 10z.$$

This forces z to be either $z = 0$, 1, or 2. These cases each reduce the problem to a problem like that in section 52.

Diophantes

54. The Life of Diophantes

Diophantes worked in the field of indeterminate equations in ancient times. Equations whose solutions are integers are called "Diophantine" in his honor. When he died his students summed up his life with a story that was also a Diophantine problem.

1/6 of his life the gods granted him for his youth.
In 1/12 he became of age.
In 1/7 he married.
In 5 years his son was born.
For 1/2 of his life his son lived.
Then in four years he died.

The puzzle, put into modern algebraic form, gives the equation

$$x/6 + x/12 + x/7 + 5 + x/2 + 4 = x$$

Clearing fractions and solving gives $x = 84$. So his youth lasted until 14; he became of age at 21; married at 33; the birth of his son took place hen he was 38; the death of his son happened when he was 80 years of age.

VIII. Sums of Squares

Pythagorean Triples

55. The Formula of Pythagoras

By the term "Pythagorean triangle" we mean a right triangle whose sides satisfy

$$x^2 + y^2 = z^2$$

and have integral values. The best known is the triangle whose sides are

$$3, \quad 4, \quad 5.$$

Other possible values for x, y, and z are:

5,	12,	13	8,	15,	17
6,	8,	10	9,	12,	15

The latter pair are not genuinely new but are derived from 3, 4, 5 by multiplying.

Pythagoras devised a formula that could be used to obtain new triples. If we give the formula as in ancient times, it must be done verbally. The modern algebraic notation is given to the left of the verbal formula in the ancient style.

n Pick any number you like.

$x = 2n + 1$ Double it and add one.

$y = 2n^2 + 2n$ Double the square and add the double number.

$z = 2n^2 + 2n + 1$ Double the square and add the double number, and then add one.

These formulas implicitly suppose that $n > 0$. We may check what results from the first few values of n.

for $n = 1$	for $n = 2$	for $n = 3$	for $n = 4$
$x = 3$	$x = 5$	$x = 7$	$x = 9$
$y = 4$	$y = 12$	$y = 24$	$y = 40$
$z = 5$	$z = 13$	$z = 25$	$z = 41$

for $n = 5$	for $n = 6$	for $n = 7$	for $n = 8$
$x = 11$	$x = 13$	$x = 15$	$x = 17$
$y = 60$	$y = 84$	$y = 112$	$y = 144$
$z = 61$	$z = 85$	$z = 113$	$z = 145$

It is easily checked that the formula is correct, that is, that it must give nubers x, y, and z that satisfy $x^2 + y^2 = z^2$.

$$(2n + 1)^2 + (2n^2 + 2n)^2 = (2n^2 + 2n + 1)^2$$

It does not, however, yield all the possible values. The triple

$$8, \quad 15, \quad 17$$

forms a right triangle but can never be obtained using the formula of Pythagoras.

56. A Modern Formula

$$x = u^2 - v^2$$

$$y = 2uv$$

$$z = u^2 + v^2$$

The conditions on these equations are: $u > 1$, $v > 0$, and $u > v$. They can be verified by substituting into

$$x^2 + y^2 = z^2$$

giving the identity

$$(u^2 - v^2)^2 + 4u^2v^2 = (u^2 + v^2)^2.$$

This modern formula allows the choice of values for u and v both. Each choice will give values of x, y, and z. For example, we may obtain Pythagorean triples as follows:

$$u = 2, v = 1 \qquad u = 4, v = 1$$
$$x = 3 \qquad\qquad x = 15$$
$$y = 4 \qquad\qquad y = 8$$
$$z = 5 \qquad\qquad z = 17$$

57. Comparing the Formulas

The formula of Pythagoras has $z = y + 1$. So we may ask what the modern formula does in this special case. If we demand that

$$\text{hypotenuse} = \text{leg} + 1$$

we have two possible cases for the modern formula:

$$x = u^2 - v^2 \qquad y = 2uv \qquad z = u^2 + v^2.$$

It must be that $z = y + 1$ or the alternative $z = x + 1$; these cases are displayed independently in two columns.

$z = y + 1$	$z = x + 1$
$u^2 + v^2 = 2uv + 1$	$u^2 + v^2 = u^2 - v^2 + 1$
$u^2 - 2uv + v^2 = 1$	$2v^2 = 1$
$(u - v)^2 = 1$	$v^2 = 1/2$
$u - v = 1$	$v = \sqrt{2}$
$u = v + 1$	impossible for integral values

So in the event that the hypotenuse is to exceed the length of a side by one, it can only be that $u = v + 1$. Next, we substitute this requirement back into the formulas for x, y, and z.

$$x = (v + 1)^2 - v^2 = 2v + 1$$

$$y = 2(v + 1)v = 2v^2 + 2v$$

$$z = (v + 1)^2 + v^2 = 2v^2 + 2v + 1$$

These formulas can be seen to agree with those of Pythagoras in the previous section.

58. Leg Plus Two

Next, let us consider what takes place when

$$\text{hypotenuse} = \text{leg} + 2.$$

Proceeding as in the previous section we have

$$u^2 + v^2 = 2uv + 2$$
$$u^2 - 2uv + v^2 = 2$$
$$(u - v)^2 = 2$$
$$u - v = \sqrt{2}$$
impossible for integers

$$u^2 + v^2 = u^2 - v^2 + 2$$
$$2v^2 = 2$$
$$v^2 = 1$$
$$v = 1$$

So we obtain $\qquad x = u^2 - 1, \;\; y = 2u, \;\; z = u^2 + 1.$

for $u = 2$	for $u = 3$	for $u = 4$	for $u = 5$
$x = 3$	$x = 8$	$x = 15$	$x = 24$
$y = 4$	$y = 6$	$y = 8$	$y = 10$
$z = 5$	$z = 10$	$z = 17$	$z = 26$

59. Leg Plus Three

Continuing on from the previous sections, let

$$\text{hypotenuse} = \text{leg} + 3.$$

$$u^2 + v^2 = 2uv + 3$$
$$u^2 - 2uv + v^2 = 3$$
$$(u - v)^2 = 3$$
$$u - v = \sqrt{3}$$
impossible for integers

$$u^2 + v^2 = u^2 - v^2 + 3$$
$$2v^2 = 3$$
impossible for integers

So there can be no Pythagorean triple whose hypotenuse is three more than one of its legs.

60. A Problem on Pythagorean Triples

Problem: Find a right triangle one leg of which is 100.

Solution: Use the equations

$$x = u^2 - v^2$$
$$y = 2uv$$
$$z = u^2 + v^2 .$$

Let $y = 100 = 2uv$. So $uv = 50$. To find the possible values taken by u and v, factor 50 into primes, this gives $50 = 2 \cdot 5^2$. So we obtain the factorizations

50	1
25	2
10	5

These choices for u and v each give values for x, y, and z.

u = 50
v = 1
x = 2499
y = 100
z = 2501

u = 25
v = 2
x = 621
y = 100
z = 629

u = 10
v = 5
x = 75
y = 100
z = 125

61. A Survey of Triples

We shall survey the Pythagorean triples whose numbers do not exceed 50 using the formulas

$$x = u^2 - v^2$$
$$y = 2uv$$
$$z = u^2 + v^2 .$$

	x	y	z
$u = 2, v = 1$	3	4	5
$u = 3, v = 1$	8	6	10
$v = 2$	5	12	13
$u = 4, v = 1$	15	8	17
$v = 2$	12	16	20
$v = 3$	7	24	25
$u = 5, v = 1$	24	10	26
$v = 2$	21	20	29
$v = 3$	16	30	34
$v = 4$	9	40	41
$u = 6, v = 1$	35	12	37
$v = 2$	32	24	40
$v = 3$	27	36	45
$v = 4$	20	48	52 ✖
$v = 5$	11	60	61 ✖
$u = 7, v = 1$	48	14	50

Two of these entries, the ones marked in the table, lead to legs of length greater than 50. There remain 14 Pythagorean triples that lie in the specified range.

Generalizations

62. Fermat's Problem

Is it possible to find triples of numbers like the Pythagorean triples, but for higher powers?

$$x^3 + y^3 = z^3 \ ,$$

$$x^4 + y^4 = z^4 \ ,$$

in general

$$x^n + y^n = z^n \ .$$

A great deal of effort was devoted to the question without actually finding a number $n > 2$ for which such a triple exists or proving that one could not exist.

[Accompanied by widespread publicity it was announced in 1993 that no triple can exist for any $n > 2$. The argument settling this celebrated problem was quite complex and there was still the possibility of an error as these notes were being edited in 1995. It was already known that there can be no such triples for various small values of n. – Ed.]

Representation as a Sum of Squares

63. Sums of Squares

We may ask which numbers can be expressed as the sums of squares. The square of an even number is $\equiv 0$ (mod 4) since

$$(2n)^2 = 4n.$$

The square of an odd number is $\equiv 1$ (mod 4) since

$$(2n + 1)^2 = 4n^2 + 4n + 1 \equiv 1 \ (\text{mod } 4)$$

So a number $4n + 3$ that is $\equiv 3$ (mod 4) can never be the sum of squares. A survey of small numbers is given below. Numbers that show a cross are not the sum of two squares.

1	✖		26 =	$1^2 + 5^2$
2 =	$1^2 + 1^2$		27	✖
3	✖		28	✖
4	✖		29 =	$2^2 + 5^2$
5 =	$1^2 + 2^2$		30	✖
6	✖		31	✖
7	✖		32 =	$4^2 + 4^2$
8 =	$2^2 + 2^2$		33	✖
9	✖		34 =	$3^2 + 5^2$
10 =	$1^2 + 3^2$		35	✖
11	✖		36	✖
12	✖		37 =	$1^2 + 6^2$

$$13 = 2^2 + 3^2 \qquad 38 \quad \text{✘}$$
$$14 \quad \text{✘} \qquad\qquad 39 \quad \text{✘}$$
$$15 \quad \text{✘} \qquad\qquad 40 = 2^2 + 6^2$$
$$16 \quad \text{✘} \qquad\qquad 41 = 4^2 + 5^2$$
$$17 = 1^2 + 4^2 \qquad 42 \quad \text{✘}$$
$$18 = 3^2 + 3^2 \qquad 43 \quad \text{✘}$$
$$19 \quad \text{✘} \qquad\qquad 44 \quad \text{✘}$$
$$20 = 2^2 + 4^2 \qquad 45 = 3^2 + 6^2$$
$$21 \quad \text{✘} \qquad\qquad 46 \quad \text{✘}$$
$$22 \quad \text{✘} \qquad\qquad 47 \quad \text{✘}$$
$$23 \quad \text{✘} \qquad\qquad 48 \quad \text{✘}$$
$$24 \quad \text{✘} \qquad\qquad 49 \quad \text{✘}$$
$$25 = 3^2 + 4^2 \qquad 50 = 5^2 + 5^2 \text{ and } 1^2 + 7^2$$

What guidelines can we find in this chaos of facts?

64. Algebraic Transformations

Problem: Write 481 as the sum of squares.

Solution: 481 factors into primes as 13·37. From the survey in the table immediately above we know that 13 and 37 can be represented as sums of squares.

$$481 = 13{\cdot}37 = (2^2 + 3^2)(1^2 + 6^2)$$

The right hand side of this equation has the form

$$(a^2 + b^2)(c^2 + d^2) = a^2c^2 + b^2c^2 + a^2d^2 + b^2d^2$$

It is readily checked that this is equivalent to either of

$$(ac + bd)^2 + (ad - bc)^2 \quad \text{or} \quad (ad + bc)^2 + (ac - bd)^2$$

In our specific case: $a = 2$, $b = 3$, $c = 1$, and $d = 6$; so we have the two representations

$$(2 + 18)^2 + (12 - 3)^2 \quad \text{or} \quad (12 + 3)^2 + (2 - 18)^2$$

$$20^2 + 9^2 \quad \text{or} \quad 15^2 + 16^2$$

So there are two distinct solutions to the problem: 481 has more than one expression as a sum of squares.

65. More Problems on Sums of Squares

Problem: Express 1105 as a sum of squares.

Solution: First, factor 1105 into primes. This is done in the familiar way by successively recognizing and removing prime divisors of 1105.

$$1105 \,\lfloor\, 5$$

$$221 \,\lfloor\, 13$$

$$17 \,\lfloor\, 17$$

$$1 \,\lfloor\,$$

This shows that $1105 = 5 \cdot 13 \cdot 17$. These three factors are themselves sums of squares as can be seen from the list on page 96-97; that is,

$$1105 = (1^2 + 2^2)(2^2 + 3^2)(1^2 + 4^2) =$$
$$[(1^2 + 2^2)(2^2 + 3^2)](1^2 + 4^2)$$

Using the methods of section 64, page 97, the product in brackets may be broken down into a sum in either one of two ways.

(i) First, use the identity

$$(a^2 + b^2)(c^2 + d^2) = (ac + bd)^2 + (ad - bc)^2$$

on the numerical expression in brackets.

$$(1^2 + 2^2)(2^2 + 3^2)](1^2 + 4^2) =$$

$$= [(2 + 6)^2 + (3 - 4)^2](1^2 + 4^2) = (8^2 + 1^2)(1^2 + 4^2);$$

(ii) Second, the alternative identity

$$(a^2 + b^2)(c^2 + d^2) = (ad + bc)^2 + (ac - bd)^2$$

may be used instead. It gives:

$[(1^2 + 2^2)(2^2 + 3^2)](1^2 + 4^2) =$

$[(3 + 4)^2 + (2 - 6)^2](1^2 + 4^2) =$

$(7^2 + 4^2)(1^2 + 4^2).$

Repeating the process, each of these products can be expressed as a sum in two ways giving four representations of the original factorization of 1105.

(ia) $(8^2 + 1^2)(1^2 + 4^2) = (8 + 4) + (32 - 1)^2{}^2 = 12^2 + 31^2;$

(ib) $(8^2 + 1^2)(1^2 + 4^2) = (32 + 1)^2 + (8 - 4)^2 = 33^2 + 4^2;$

(iia) $(7^2 + 4^2)(1^2 + 4^2) = (7 + 16)^2 + (28 - 4)^2 = 23^2 + 24^2;$

(iib) $(7^2 + 4^2)(1^2 + 4^2) = (28 + 4)^2 + (7 - 16)^2 = 32^2 + 9^2.$

So 1105 has been represented as a sum of squares in four ways.

Problem: Which of the following can be expressed as a sum of squares?
    (i) 101   (ii) 234   (iii) 365   (iv) 1947

Solution: We see immediately that the first of these problems, 101, satisfies $101 = 10^2 + 1^2$; so (i) can be represented as a sum of squares.
    Turning next to (ii): factor 234 into primes.

$234 \lfloor 2$

$117 \lfloor 3$ $13 \lfloor 13$

$39 \lfloor 3$ $1 \lfloor$

Two of these factors, 2 and 13, can be expressed as a sum of primes; but 3 cannot. So the methods of section 64 do not help here. We may check to see if there is reason to believe that 234 is not a sum of squares. Since $234 \equiv 3 \pmod 4$, it cannot be represented as a sum of squares as pointed out in section 63, page 96. This finishes (ii).

Now turn to (iii). $365 = 5 \cdot 73$ and both of these factors are sums of squares, thus 365 is too.

For (iv), $1947 \equiv 3 \pmod 4$ and so it cannot be a sum of squares.

Problem: Express 2250 as a sum of squares.

Solution: Begin by factoring 2250.

$$2250 \,\lfloor\, 2$$

$$1125 \,\lfloor\, 3$$

$$375 \,\lfloor\, 3$$

$$125 \,\lfloor\, 5$$

$$25 \,\lfloor\, 5$$

$$5 \,\lfloor\, 5$$

$$1 \,\lfloor$$

$$2250 = 2 \cdot 3^2 \cdot 5^3 = (2 \cdot 5^2)(3^2 \cdot 5) = 50 \cdot 45 = (7^2 + 1^2)(6^2 + 3^2) \text{ or}$$

$$(5^2 + 5^2)(6^2 + 3^2)$$

case ia) $(7^2 + 1^2)(6^2 + 3^2) = (42 + 3)^2 + (21 - 6)^2$
$$= 45^2 + 15^2;$$

case ib) $(7^2 + 1^2)(6^2 + 3^2) = (21 + 6)^2 + (42 - 3)^2$
$$= 27^2 + 39^2;$$

case iia) $(5^2 + 5^2)(6^2 + 3^2) = (30 + 15)^2 + (15 - 30)^2 =$
$$45^2 + 15^2;$$

case iib) $(5^2 + 5^2)(6^2 + 3^2) = (15 + 30)^2 + (30 - 15)^2 =$
$$45^2 + 15^2.$$

The four cases give only two distinct representations, so we have
$$2250 = 15^2 + 45^2 = 27^2 + 39^2.$$

Problem: Express 1394 as a sum of squares.

Solution: Factoring gives

$$1394 \mid 2$$
$$697 \mid 17$$
$$41 \mid 41$$
$$1 \mid$$

So $1394 = 2 \cdot 17 \cdot 41$. Our survey on page 96-97 has shown that all three of these prime factors are representable as a sum of squares. In fact, we have the following relationships appearing among the divisors of 1394:

$$17 = 4^2 + 1^2;$$
$$34 = 5^2 + 3^2;$$
$$41 = 5^2 + 4^2;$$
$$82 = 9^2 + 1^2.$$

In the cases labeled "a" below, use the identity

$$(a^2 + b^2)(c^2 + d^2) = (ac + bd)^2 + (ad - bc)^2$$

and in the cases labeled "b" below, use the alternative identity:

$$(a^2 + b^2)(c^2 + d^2) = (ad + bc)^2 + (ac - bd)^2.$$

case ia) $1394 = 17 \cdot 82 = (4^2 + 1^2)(9^2 + 1^2) =$
$$(36 + 1)^2 + (4 - 9)^2 = 37^2 + 5^2;$$

case ib) $1394 = 17 \cdot 82 = (4^2 + 1^2)(9^2 + 1^2) =$
$$(4 + 9)^2 + (36 - 1)^2 = 13^2 + 35^2;$$

case iia) $1394 = 34 \cdot 41 = (5^2 + 3^2)(5^2 + 4^2) =$
$$= (25 + 12)^2 + (20 - 15)^2 = 35^2 + 13^2;$$

case iib) $1394 = 34 \cdot 41 = (5^2 + 3^2)(5^2 + 4^2)$
$$= (20 + 15)^2 + (25 - 12)^2 = 35^2 + 13^2;$$

As a result we have: $1394 = 35^2 + 13^2 = 37^2 + 5^2$.

Problem: Express the product of 32, 37, and 41 as a sum of squares.

Solution: $32 \cdot 37 \cdot 41 = (4^2 + 4^2)(6^2 + 1^2)(5^2 + 4^2)$

$$(4^2 + 4^2)(6^2 + 1^2) = (24 + 4)^2 + (4 - 24)^2 = 28^2 + 20^2$$
$$\text{and} = (24 - 4)^2 + (4 + 24)^2 = 20^2 + 28^2;$$

$$(20^2 + 28^2)(5^2 + 4^2) = (100 + 112)^2 + (80 - 140)^2 =$$
$$= 212^2 + 60^2 \quad \text{and} \quad = (100 - 112)^2 + (80 + 140)^2$$
$$= 12^2 + 220^2.$$

So $32 \cdot 37 \cdot 41 = 212^2 + 60^2 = 12^2 + 220^2$.

66. Four Squares

Although there are many numbers that are not the sum of two squares, it is a fact that every positive integer can be expressed as a sum of four squares.

67. Primes

Primes of the form $4n + 1$ can be expressed as a sum of two squares. This can easily be checked for the smaller examples.

$$5 = 2^2 + 1^2$$
$$13 = 3^2 + 2^2$$
$$17 = 4^2 + 1^2$$
$$29 = 5^2 + 2^2$$
$$37 = 6^2 + 1^2$$
$$41 = 5^2 + 4^2$$
$$53 = 7^2 + 2^2$$
$$61 = 6^2 + 5^2$$
$$73 = 8^2 + 3^2$$
$$89 = 8^2 + 5^2$$
$$97 = 9^2 + 4^2$$

68. Repeated Factors

Let a number n be a sum of two squares, say, $n = 4^2 + 5^2$. A number of the form $a^2 n$ can be expressed as a sum of two squares also because

$$a^2 n = (4^2 + 5^2)a^2 = (4a)^2 + (5a)^2.$$

To see how this is used, let us compare the numbers 1947 and 234.

When we factor 1947, we obtain

$$1947 \lfloor\ 3 \qquad\qquad (4n + 3)$$

$$649 \lfloor\ 11 \qquad\qquad (4n + 3)$$

$$59 \lfloor\ 59 \qquad\qquad (4n + 3)$$

The fact that any one of these factors is $\equiv 3 \pmod 4$ is sufficient to guarantee that 1947 cannot be expressed as a sum of squares.

Next, turn to 234.

$$234 \lfloor\ 2$$

$$117 \lfloor\ 3 \qquad\qquad (4n + 3)$$

$$39 \lfloor\ 3 \qquad\qquad (4n + 3)$$

$$13 \lfloor\ 13$$

The factor $3 \equiv 3 \pmod 4$ but it appears twice, so we may use the technique given at the beginning of this section and write

$$234 = 26 \cdot 3^2 = (5^2 + 1^2) \cdot 3^2 = 15^2 + 3^2.$$

69. Additional Examples

(i) 1666

$$1666 \,\lfloor\, 2$$
$$833 \,\lfloor\, 7 \qquad 34 = 5^2 + 3^2$$
$$119 \,\lfloor\, 7$$
$$17 \,\lfloor\, 17 \qquad 1666 = 7^2(5^2 + 3^2) = 35^2 + 21^2$$
$$1 \,\lfloor$$

(ii) 584

$$584 \,\lfloor\, 2$$
$$292 \,\lfloor\, 2 \qquad (8^2 + 3^2)(1^2 + 1^2)2^2 =$$
$$146 \,\lfloor\, 2 \qquad [(8 + 3)^2 + (8 - 3)^2]2^2 = (11^2 + 5^2)2^2 =$$
$$73 \,\lfloor\, 73 \qquad\qquad 22^2 + 10^2$$
$$1 \,\lfloor$$

(iii) 850

$$850 \,\lfloor\, 2$$
$$425 \,\lfloor\, 5 \qquad 34 = 5^2 + 3^2$$
$$85 \,\lfloor\, 5 \qquad 850 = 25^2 + 15^2$$
$$17 \,\lfloor\, 17$$
$$1 \,\lfloor$$

(iv) 6208

$6208 \lfloor 2$ $776 \lfloor 2$

$3104 \lfloor 2$ $388 \lfloor 2$

$1552 \lfloor 2$ $194 \lfloor 2$ $97 \lfloor 97$

Here the computation has been broken into several columns instead of being written in one column.

So we have that $6208 = 2^{6} \cdot 97 = 2^{6}(9^2 + 4^2) = 8^2(9^2 + 4^2) = 72^2 + 32^2$.

(v) 27,521

$27{,}521 \lfloor 13 = 3^2 + 2^2$

$2117 \lfloor 29 = 5^2 + 2^2$

$73 \lfloor 73 = 8^2 + 3^2$

$$(3^2 + 2^2)(5^2 + 2^2) = (15 + 4)^2 + (6 - 10)^2 = 19^2 + 4^2$$
$$= (15 - 4)^2 + (6 + 10)^2 = 11^2 + 16^2$$

$$(19^2 + 4^2)(8^2 + 3^2) = (152 + 12)^2 + (57 - 32)^2 = 164^2 + 25^2 =$$
$$164^2 + 25^2 = (152 - 12)^2 + (57 + 32)^2 = 140^2 + 89^2$$

$$27{,}251 = (11^2 + 16^2)(8^2 + 3^2) =$$
$$= (88 + 48)^2 + (33 - 128)^2 = 136^2 + 95^2 =$$
$$= (88 - 48)^2 + (33 + 128)^2 = 40^2 + 161^2$$

70. Euler's Method of Factoring

Euler took tables of sums of squares and used them for factoring.

$$235^2 + 972^2 = 1{,}000^2 + 3^2 = 1{,}000{,}009$$

This has the form $N = a^2 + b^2 = c^2 + d^2$

so $a^2 - c^2 = d^2 - b^2$

$$(a + c)(a - c) = (d + b)(d - b)$$

Returning to the numerical values

$$238 \cdot 232 = 1972 \cdot 28$$

$$(7 \cdot 34)(58 \cdot 4) = (58 \cdot 34)(7 \cdot 4)$$

These last factorizations can be obtained by examining the numbers 28, 232, 238, and 1972.

28 $\lfloor$ 2	232 $\lfloor$ 2	238 $\lfloor$ 2	1972 $\lfloor$ 2
14 $\lfloor$ 2	116 $\lfloor$ 2	119 $\lfloor$ 7	986 $\lfloor$ 2
7 $\lfloor$ 7	58 $\lfloor$ 2	17 $\lfloor$ 17	493 $\lfloor$ 17
	29 $\lfloor$ 29		29 $\lfloor$ 29

$$28 = 2^2 \cdot 7$$
$$232 = 2^3 \cdot 29$$
$$238 = 2 \cdot 7 \cdot 17$$
$$1972 = 2^2 \cdot 17 \cdot 29$$

$$1,000,009 = (2^2 + 17^2)(7^2 + 58^2) = 293 \cdot 3413$$

So beginning with the fact that 1,000,009 can be expressed as a sum of squares in two different ways, he was able to factor it in spite of its size.

IX. Sequences

71. Square Numbers

The sequence of squares:

0   1   4   9   16   25   36   49   64   81   100

has as its sequence of differences between successive terms

1   3   5   7   9   11   13   15   17   19.

That is to say, any odd number can be expressed as the difference of successive squares:

$$7 = 4^2 - 3^2, \qquad 9 = 5^2 - 4^2, \qquad 17 = 9^2 - 8^2.$$

72. Cubes

Beginning with the cubes we may form the sequence of differences, the sequence of differences of these, and so on. They are placed here so that each entry is the difference between the numbers immediately above it.

```
0   1   8   27   64   125   216   343   512   729   1000
  1   7   19   37   61   91   127   169   217   271
    6   12   18   24   30   36   42   48   54
      6   6   6   6   6   6   6   6
        0   0   0   0   0   0   0
```

73. Fourth Powers

When the fourth powers are treated in this same way, an extra line is needed before reaching a row of 0's.

```
0   1   16   81   256   625   1296   2401   4096   6561   10000
  1   15   65   175   369   671   1105   1695   2465   3439
    14   50   110   194   302   434   590   770   974
      36   60   84   108   132   156   180   204
        24   24   24   24   24   24   24
          0   0   0   0   0   0
```

74. The General Pattern

With squares the sequence of second differences is constantly 2; with cubes the sequence of third differences is constantly 6; with fourth powers the fourth differences become constantly 24.

$$2! = 2 \qquad = 1 \cdot 2$$
$$3! = 6 \qquad = 1 \cdot 2 \cdot 3$$
$$4! = 24 \qquad = 1 \cdot 2 \cdot 3 \cdot 4$$
$$5! = 120 \quad = 1 \cdot 2 \cdot 3 \cdot 4 \cdot 5$$

A similar situation arises in taking successive derivatives.

If $y = x^2$, then $\dfrac{dy}{dx} = 2x$ and $\dfrac{d^2y}{d^2x} = 2$.

If $y = x^3$, then $\dfrac{dy}{dx} = 3x^2$, $\dfrac{d^2y}{d^2x} = 6x$, and $\dfrac{d^3y}{d^3x} = 6$.

And in general if $y = x^n$, then $\dfrac{d^ny}{d^nx} = n!$

75. Variable Powers

If we form the successive powers of a number, for instance $3, 3^2, 3^3, 3^4, 3^5, \cdots$, we find the differences no longer simplify to a constant.

```
1   3   9   27   81   243   729   2187   6561   19683   59049
  2   6   18   54   162   486   1458   4374   13122   39366
    4   12   36   108   324   972   2916   8748   26244
      8   24   etc.
      16        etc.
```

When the sequence is the powers of two, the same sequences of differences repeat to infinity. Two is the only number whose powers reproduce themselves in this way.

1   2   4   8   16   32   64   128   256   512   1024
  1   2   4   8   16   32   64   128   256   512
    1   2   4   8   16   32   64   128   256
          etc.

[The lecture course ended with this prelude to seequences. It is obvious from the way the notes ended that von Baravalle had developed a much more extensive treatment of the subject.]

The Number π

Two outstanding constants of mathematics have been dealt with in previous articles in *The Mathematics Teacher*; the number e, the base of natural logarithms, and the number G, the ratio of the Golden Section. [Dec. 1945 issue, and Jan. 1948.] To complete this series, the present article takes up the third and best known constant, the number π.

As its symbol indicates (π stands for periphery), it represents the ratio of the two outstanding dimensions of the circle, the way around it and the distance across it;

$$\pi = \frac{\text{circumference of a circle}}{\text{diameter of the circle}}.$$

Expressing the diameter in terms of its radius r, we obtain the formula for the circumference of the circle c:

$$\frac{c}{2r} = \pi ; \qquad c = 2\pi r.$$

This is by far not the only ratio in which this constant appears. For example π is also the ratio of the area of a circle A to the area of the square erected on its radius r:

$$\pi = \frac{A}{r^2} ; \qquad A = r^2\pi .$$

It further appears in many other formulae. The volume (V) of a circular cylinder with a base radius r and altitude h is

$$V = r^2\pi h$$

and of a circular cone –

$$V = \frac{r^2\pi h}{3} .$$

The surface of a sphere is

$$A = 4r^2\pi$$

and its volume –

$$V = \frac{4}{3}r^3\pi.$$

The domain of π also extends beyond circular structures. The area of an ellipse with semi-axes a and b is

$$A = ab\pi,$$

and the volume of an ellipsoid with the three semi-axes a, b and c is

$$V = \frac{4}{3}abc\pi.$$

The area of a cardioid drawn in Figure 1 as an envelope of circles is

$$A = \frac{3}{2}a^2\pi.$$

in which a stands for the diameter of the circle, whose circumference is indicated by the dotted line.

Figure 1

To construct Figure 1, the dotted circle is divided into thirty-two equal parts. Each of the thirty-two points of division becomes the center of a circle whose radius is the distance from the highest point on the circle (upper end of the vertical diameter).

Further examples of curves whose formulae contain π are the roses. The area enclosed by the three-leafed rose (black portions in Figure 2) is

Figure 2

$$A = \frac{1}{4}a^2\pi,$$

in which a stands for the area of the circle circumscribed around it. The area enclosed by a four-leafed rose (black area in Figure 3) is

$$A = \frac{1}{2}a^2\pi$$

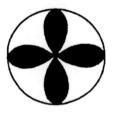

Figure 3

in which a again denotes the radius of the circumscribed circle. The volume of a ring (torus), obtained by rotating a circle with radius a about an axis in the same plane at a distance of b

units from the center of the circle is expressed in the following formula:

$$V = 2a^2b\pi^2.$$

The volume of the solid of rotation produced by rotating an astroid, Figure 4, about one of its axes is

$$V = \frac{32}{105}a^3\pi,$$

Here a represents the distance of any of the star points from the center, the radius of the circumscribed circle. The surface area of the same solid of rotation is

$$A = \frac{12}{5}a^2\pi.$$

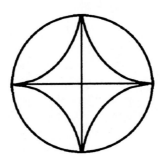

Figure 4

The formula for the volume of the solid

$$x^{2/3} + y^{2/3} + z^{2/3} = a^{2/3},$$

whose traces on the coordinate planes are astroids, also contains π:

$$V = \frac{4}{35}a^3\pi.$$

All these formulas are obtained by integral calculus.

We can go beyond areas, surfaces, and volumes to find π again in a variety of relationships. A semicircle (radius r), cut out of sheet metal and balanced on a point of support lies on its axis of symmetry at a distance d from the center, which is

$$d = \frac{4r}{3\pi}.$$

π even appears in formulae of probability, statistics, and in the field of an actuary.

Any vibration, mechanical, acoustical, or electrical, proceeds with varying speed. By determining the distance covered by a point on a vibrating musical chord between its extreme positions during a certain time unit, we obtain its average speed of motion. The actual speed of the point is greater every time the chord is near to passing its middle position. It is less than the average speed every time the point finds itself near to one of its extreme elongations. The maximum speed occurs when the point passes in either direction through its position of rest. This maximum speed is in any vibration exactly $\pi/2$ times the average speed. As this holds good for vibrations accompanying every sound of our own vocal chords and in the air around us, π is contained in every word and sentence we say.

The differential equation of a vibration is

$$\frac{d^2x}{dt^2} = -a^2x,$$

and its complete solution is $x = c \sin(at - \alpha)$. In this equation c and a represent arbitrary constants. For $x = 0$ and $t = 0$, the solution is

$$x = c \sin at.$$

Maximum speed:

$$\frac{dx}{dt} = ac \cos at; \qquad \text{for } t = 0: \left.\frac{dx}{dt}\right|_{max} = a{\cdot}c.$$

For $at = \pi/2$; $x = c$ and $\dfrac{x}{t} = \dfrac{2ac}{\pi}$,   The ratio $\dfrac{\dfrac{dx}{dt}\Big|_{max}}{\dfrac{x}{t}}$ is

therefore $\dfrac{\pi}{2}$.

The value of π up to 22 decimal places is

$$3.1415926535897932384626\cdots.$$

These successive numerals are the same as the number of letters contained in the successive words of the French verse:

> *"Que j'aime a faire apprendre*
> *Un nombre utile aux sages*
> *Immortel Archimède, artiste ingénieur*
> *Qui de ton judgement pent priser la valeur."*

Translation: "How I like to teach a number, useful to the learned: Immortal Archimedes, skillful investigator, yes, the number can tell the praise of your judgment."

Until recently π had been calculated to 707 decimal places. This figure had been obtained by an Englishman, William Shanks, in 1853. With the help of modern electronic computing machines, the number of decimal places has now been extended to over 2,000.

The history of π dates back 3,500 years, as far as historical records show. The Egyptian Rhind Papyrus, dating as far back as 1700 B.C., gives directions for obtaining the area of a circle. Expressed in modern symbols, its formula with A for the circle's area and d for its diameter, is as follows:

$$A = \left(d - \frac{1}{9}d\right)^2 = d^2\left(1 - \frac{1}{9}\right)^2 = 4r^2\left(\frac{8}{9}\right)^2 = r^2\frac{4\cdot 64}{81} = r^2\frac{256}{81}$$

The fraction 256/81, which here takes the place of π, equals in decimals $3.16050\cdots$. Compared with π, ($3.14159\cdots$), the difference is $0.01891\cdots$, or less than 1/50.

Archimedes expresses π numerically as follows:

$$3\frac{1}{7} > \pi > 3\frac{10}{71}$$

Expressed in decimals the same relationship would read

$$3.142857\cdots > \pi > 3.140845\cdots .$$

Midway between these two values of Archimedes lies the number 3.141851, which is only 0.000259··· or about 2½ ten-thousands greater than π. In ancient China, π was expressed by Ch'ang Höng (125 A.D.) as 3.162···, the accuracy of which is only slightly less than the value given in the Egyptian papyrus. In 265 A.D. Wang Fan expressed the value of π by the fraction 142/45, or 3.15555···. In 470 A.D. Ch'ung-chih gave a different fraction: 355/113, or 3.1415929···, which is correct all the way out to 6 decimal places.   In India Aryabhata (510 A.D.) expressed π this way: "Add 4 to 100, multiply by 8 and add 62,000.   This is the approximate circumference of a circle whose diameter is 20,000." Thus π appears as the fraction 62832/20000, which resolves to 3.1416, and is less than one ten thousandth off.

Though some of these values are sufficiently accurate to have met the practical demands of their times, none reveals any mathematical regularity for the value of π. Against the background of the philosophers of antiquity, one can appreciate the great disappointment this fact caused to mathematicians and philosophers. Failure of the outstanding ratio of the most perfect curve to conform to any pattern of mathematical regularity was considered a blemish upon the divine world order and never accepted as the ultimate answer.

The anticipations of antiquity regarding π finally proved justified, but the solution was found only as recently as 360 years ago. The value of π was first expressed in a regular mathematical pattern in 1592 by the great French mathematician, François Viète (1540 - 1603), who found:

$$\pi = 2 \frac{1}{\sqrt{1/2} \cdot \sqrt{1/2 + 1/2\sqrt{1/2}} \cdot \sqrt{1/2 + 1/2\sqrt{1/2 + 1/2\sqrt{1/2}} \cdots}}$$

The denominator is an infinite product of expressions of square roots with a regular structure. The possibility of one such development suggests the possibility of other simpler ones; actually, in 1655, John Wallis, an English mathematician (1616-1703), found

$$\pi = 4 \, \frac{2 \cdot 4 \cdot 4 \cdot 6 \cdot 6 \cdot 8 \cdot 8 \cdot 10 \cdot 10 \cdot 12 \cdot 12 \cdots}{3 \cdot 3 \cdot 5 \cdot 5 \cdot 7 \cdot 7 \cdot 9 \cdot 9 \cdot 11 \cdot 11 \cdot 13 \cdots}.$$

Here π is expressed by infinite products of numbers, appearing in both numerator and denominator of a fraction but without any roots. In the numerator we find the even numbers, in the denominator the odd numbers. Both appear in pairs with the exception of the first factor in the numerator. Only three years later, in 1658, William Brouncker (1620 - 1684) expressed the value of π as a continued fraction:

$$\pi = 4 \cdot \cfrac{1}{1 + \cfrac{1^2}{2 + \cfrac{3^2}{2 + \cfrac{5^2}{2 + \cfrac{7^2}{2 + \cfrac{9^2}{2 + \cdots}}}}}}$$

which again shows complete regularity, the only varying figures being the squares of odd numbers.

Progress was on the march. The same century brought the final presentation of π as the limit of an infinite series of simple fractions using the odd numbers as their denominators and having alternating signs, the Leibnitz Series. The regularity which was impossible in decimal expansions of the value of π now became possible through an infinite series of common fractions. Actually, this expression in fractions was more in keeping with the work of the thinkers of antiquity than was the expression in decimals, which have been in use only since the sixteenth century. That the series is infinite (the transcendence of π was proved by F. Lindemann in 1882) makes the result even more dynamic.

The Leibnitz Series is a fruit of the calculus obtained by one of its inventors. It is derived by expanding the arctangent function according to Maclaurin's series.

$$f(x) = f(0) + \frac{f'(0)}{1!}x + \frac{f''(0)}{2!}x^2 + \frac{f'''(0)}{3!}x^3 + \cdots + \frac{f^{(n)}(0)}{n!}x^n + \cdots.$$

The form for arctan x thus reads: Arctan $x =$

$$x - \frac{x^3}{3} + \frac{x^5}{5} - \frac{x^7}{7} + \frac{x^9}{9} - \frac{x^{11}}{11} + \cdots + (-1)^{n-1}\frac{x^{2n-1}}{2n-1} + \cdots,$$

which converges for all values of x within the limits

$$-1 \leq x \leq 1.$$

Substituting $x = 1$ for an angle of 45° (in radians 45° = $\pi/4$; tan 45° = 1) we obtain:

$$\frac{\pi}{4} = 1 - \frac{1}{3} + \frac{1}{5} - \frac{1}{7} + \frac{1}{9} - \frac{1}{11} + \frac{1}{13} - \frac{1}{15} + \cdots$$

or $$\pi = 4 \cdot \left(1 - \frac{1}{3} + \frac{1}{5} - \frac{1}{7} + \frac{1}{9} - \frac{1}{11} + \frac{1}{13} - \frac{1}{15} + \cdots \right).$$

The Leibnitz Series has not been surpassed in all of subsequent history in point of its outstanding simplicity. The only later additions were devices for calculating larger numbers of decimals with less effort in the process of computation, in other words, by finding methods of developing π through faster convergence. By expanding the arcsine in the same way we obtain the formula   Arcsine $x =$

$$x + \frac{1}{2} \cdot \frac{1}{3}x^3 + \frac{1 \cdot 3}{2 \cdot 4} \cdot \frac{1}{5}x^5 + \frac{1 \cdot 3 \cdot 5}{2 \cdot 4 \cdot 6} \cdot \frac{1}{7}x^7 + \frac{1 \cdot 3 \cdot 5 \cdot 7}{2 \cdot 4 \cdot 6 \cdot 8} \cdot \frac{1}{9}x^9 + \cdots$$

which covers all values of x within the limits of $-1 \le x \le 1$. Substituting $x = 1$, we obtain for arcsin 1, corresponding to an angle of 90, or in radians $\pi/2$, the formula

$$\frac{\pi}{4} = 1 + \frac{1}{2} \cdot \frac{1}{3} + \frac{1 \cdot 3}{2 \cdot 4} \cdot \frac{1}{5} + \frac{1 \cdot 3 \cdot 5}{2 \cdot 4 \cdot 6} \cdot \frac{1}{7} + \frac{1 \cdot 3 \cdot 5 \cdot 7}{2 \cdot 4 \cdot 6 \cdot 8} \cdot \frac{1}{9} + \cdots$$

a series which, though more complicated than the Leibnitz Series, converges more rapidly. Further series show a still greater convergence , for instance, that which Abraham Sharp used in 1717 to calculate the value of π to 72 decimal places:

$$\pi = 6 \cdot \frac{1}{\sqrt{3}} \cdot \left(1 - \frac{1}{3 \cdot 3} + \frac{1}{3^2 \cdot 5} - \frac{1}{3^3 \cdot 7} + \frac{1}{3^4 \cdot 9} - \frac{1}{3^5 \cdot 11} \cdots \right).$$

To find the value of π geometrically, Deinostratus (350 B.C.) used a curve called the Quadratrix. Its construction is shown in Figure 5. Above and below a horizontal base AB, a

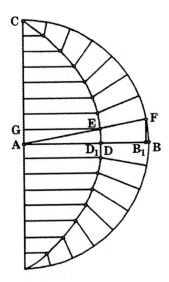

Figure 5. Geometric construction of the value of π

quarter of a circle with A as its center and AB as its radius is drawn and divided into equal parts. In Figure 5 there are 8 equal parts above and 8 below the base. Then the perpendicular radii are divided into the same number of equal parts as the quarters of the circles and through every point of the division a horizontal line is drawn.

After adding a radius through each point of the division on the circle, we start with the highest point C and mark the point where the next horizontal line and the next radius intersect, and then continue marking the intersection points of the second horizontal line and the second radius and so forth. The curve passing through these intersection points is the Quadratix. The point at which it cuts the base AB is the point D and the ratio of the line segments AB and AD is

$$\frac{AB}{AD} = \frac{\pi}{2}.$$

The length of the arc BC, one quarter of the circumference of a circle, is

$$\frac{2r\pi}{4} = r\frac{\pi}{2}.$$

The ratio of arc BC to the radius r is therefore $\pi/2$. The ratio of one-eighth of the arc BC to one-eighth of the radius is therefore also $\pi/2$. The length of the perpendicular from E to AB equals $ED_1 = AG$, which is by construction one-eighth of the radius AC; BF is ½ of BC. Therefore the ratio BF to ED_1 is still $\pi/2$. What holds true for the eighths holds true for any other fraction. The smaller each part of arc BC becomes, the closer it approaches the length of the perpendicular FB_1. Through the similarity of the triangles $\triangle AB_1F$ and $\triangle AD_1E$ we obtain the proportion

$$\frac{AB_1}{AD_1} = \frac{FB_1}{ED_1}.$$

With an increasing number of points of division and a decreasing angle FAB, B_1 approaches B, D_1 approaches D, and the ratio FB_1/ED_1 equals $\pi/2$.

The geometric aspect of π lead to the famous problem of the quadrature of the circle, the task of constructing a square (quadratum) whose area equals the area of a given circle. The curve in Figure 5 also derives its name from this problem. Archimedes made an outstanding contribution to the quadrature of the circle when he found that the area of a circle equals the area of a right triangle, one of whose legs equals the radius and the other the circumference of the circle. This discovery established an equality between the curved area of a circle and the area of a form bounded only by straight lines and made possible the construction of the quadrature of a circle immediately upon straightening out its circumference. The latter task, so easily performed every time a wheel rolls over a road imprinting its exact circumference with each revolution, has nonetheless been an age-long challenge to masters of geometrical construction. Its complete solution is possible only with the use of higher curves. Numerous approximations of this geometric construction have been found, however, which for practical purposes represent a solution. Figure 6 shows the approximation constructed by Kochansky. Two tangents are drawn through the end points of the vertical diameter to the

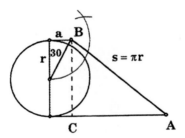

Figure 6

circle. On each of these tangents a certain point is marked. On the lower tangent the point A that has been marked is three times the length of each of the radii. On the upper tangent line the point B is fixed at the intersection of that tangent with the prolonged radius that is drawn at an angle of 30° to the vertical diameter. The distance AB is then taken to be π times the radius.

AB is the hypotenuse of the right triangle ABC. Its vertical leg is $2r$, and its horizontal leg $3r$ minus the distance a, which is one-half the length of the base of an equilateral triangle with the altitude r (that is, $a = r/\sqrt{3}$):

$$AB = \sqrt{(2r)^2 + (\frac{3r - r}{\sqrt{3}})^2} = r\sqrt{4 + \frac{(3\sqrt{3} - 1)^2}{3}} = r \cdot 3.14153.$$

The approximation provides a difference of less than 0.0001, which lies beyond the graphical limit of precision of Figure 6.

    With the help of Kochansky's construction, it is possible to effect the quadrature of the circle, as shown in Figure 7, in two steps.

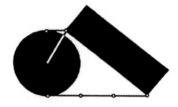

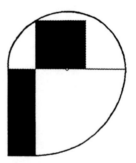

Figure 7

In the upper diagram of Figure 7 we recognize Kochansky's construction. The resulting distance is used as the base of a rectangle with an altitude equal to the radius of the circle. According to Archimedes, the area of a circle equals the area of a triangle whose base is the circumference of the circle and whose altitude is the radius. Therefore, it also equals a rectangle whose base is half the circumference of the circle and whose altitude is the radius. The next step consists of transforming the area of the rectangle into a square - a step as indicated in Figure 8. The rectangle *ADEF* is the same one in Figure 7. By construction, *DB* is equal to *DE* and the intersection of the semicircle above AB with the prolongation of *DE* determines the point *C*. $\triangle ABC$ is a right triangle with altitude h. The area of a square with h as its side equals the area of the rectangle *ADEF*.

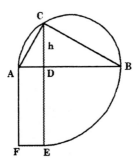

Figure 8

$$AD = \frac{h}{\tan \angle CAD};$$

$DE = DB = h \tan\angle BCD$; $\angle BCD = \angle CAD$ (angles whose sides are perpendicular). Therefore

$$AD \cdot DE = \frac{h}{\tan \angle CAD} \cdot h \tan \angle CAD = h^2.$$

The four areas which are marked in black in Figure 7 are equal to one another and show in their sequence from left to right the completion of the quadrature of the circle.

Finally, let us compare the three great constants of mathematics: G, e, π,

$$G = 0.6180339887\cdots$$

$$e = 2.7182818284\cdots$$

$$\pi = 3.1415926535\cdots$$

In the form of continued fractions

$$G = \; + \cfrac{1}{1 + \cfrac{1}{1 + \cfrac{1}{1 + \cfrac{1}{1 + \cfrac{1}{1 + \cdots}}}}}$$

$$e = \; + 1 + \cfrac{1}{1 + \cfrac{1}{2 + \cfrac{1}{1 + \cfrac{1}{4 + \cdots}}}}$$

$$\pi = 4 \cdot \cfrac{1}{1 + \cfrac{1^2}{2 + \cfrac{3^2}{2 + \cfrac{5^2}{2 + \cfrac{7^2}{2 + \cfrac{9^2}{2 + \cdots}}}}}}$$

G expresses itself through repetitions of the number 1 only and e through repetitions of 1 and the powers of 2. In the fraction of π the variable element is the series of the squares of the odd numbers.

There is an approximation between π and G which played a major role in the history of the investigations of the proportions of the Great Pyramid in Egypt:

$$(\pi/4)^2 = 0.6168\cdots$$

$$G = 0.6180\cdots$$

Though the difference between the two values is only 0.0012, their closeness is merely incidental and has no basis in mathematical law. Neither is their any mathematical connection between G and e.

In contrast, there is a distinct mathematical relationship between the constants π and e. The expansion of e^x, according to MacLaurin's Series is

$$e^x = 1 + x + \frac{x^2}{2!} + \frac{x^3}{3!} + \frac{x^4}{4!} + \frac{x^5}{5!} + \cdots + \frac{x^n}{n!} + \cdots$$

There sine and cosine are given by

$$\sin x = x - \frac{x^3}{3!} + \frac{x^5}{5!} - \frac{x^7}{7!} + \frac{x^9}{9!} + \cdots + (-1)^{n-1} \frac{x^{2n-1}}{(2n-1)!} + \cdots$$

$$\cos x = 1 - \frac{x^2}{2!} + \frac{x^4}{4!} - \frac{x^6}{6!} + \frac{x^8}{8!} + \cdots + (-1)^{n-1} \frac{x^{2n-2}}{(2n-2)!} + \cdots .$$

The two series for $\sin x$ and $\cos x$ together furnish all the terms of the series e^x, with only a discrepancy in their signs, a difficulty which does not exist for the hyperbolic sines and cosines. Their expansions have only positive terms:

$$\sinh x = x + \frac{x^3}{3!} + \frac{x^5}{5!} + \frac{x^7}{7!} + \frac{x^9}{9!} + \cdots + \frac{x^{2n-1}}{(2n-1)!} + \cdots$$

$$\cosh x = 1 + \frac{x^2}{2!} + \frac{x^4}{4!} + \frac{x^6}{6!} + \frac{x^8}{8!} + \cdots + \frac{x^{2n-2}}{(2n-2)!} + \cdots .$$

Therefore it readily appears that $e^x = \sinh x + \cosh x$. An analogous result for the trigonometric functions can be obtained if we substitute for x its product with the imaginary unit:

$$e^{ix} = 1 + ix - \frac{x^2}{2!} - i\frac{x^3}{3!} + \frac{x^4}{4!} + i\frac{x^5}{5!} - \frac{x^6}{6!} - i\frac{x^7}{7!} + \cdots$$

and separate the real and imaginary terms. Thus, the result is

$$e^{ix} = 1 - \frac{x^2}{2!} + \frac{x^4}{4!} - \frac{x^6}{6!} + \cdots + i\left(x - \frac{x^3}{3!} + \frac{x^5}{5!} - \frac{x^7}{7!} + \cdots \right)$$

or

$$e^{ix} = \cos x + i \sin x,$$

a formula which finds wide application in the solving of differential equations, particularly those connected with all types of vibrations. Substituting $x = \pi$. we obtain

$$e^{i\pi} = \cos \pi + i \sin \pi = -1 + i\cdot 0 = -1$$

which results in the formula

$$e^{\pi i} = -1.$$

It is this formula which David Eugene Smith used it in the mathematical credo placed in his library:

The Science Venerable

Voltaire once remarked , "One merit of poetry few will deny; it says more and with fewer words than prose." With equal significance we may say, "One merit of mathematica few will deny; it says more and in fewer words than any other science." The formula $e^{\pi i} = -1$ expresses a world of thought, of truth, of poetry and of religious spirit, for God eternally geometrizes."

Bibliography

Ball, W. W. Rouse, *Mathematical Recreations and Essays.* revised by H.S.M. Coxeter. Macmillan. New York. 1939.

Baravalle, Hermann von, *Die Geometry des Pentagramms und der Goldene Schnitt.* J. Ch. Mellinger Verlag. Stuttgart. 1950.

Bindel, Ernst, *Die Aegyptischen Pyramiden.* Verlag Freie Waldorfschule. Stuttgart. 1932.

De Morgan, Agustus, *A Budget of Paradoxes,* II Volume. Open Court Publishing. Chicago. 1915.

Ghyka, Matila C. *Esthétique des Propositions dans la Nature et dans les Arts.* Librairie Gallimard. Paris. 1927.

Granville, W. A., P. F. Smith, and W. R. Longley, *Elements of Differential and Integral Calculus.* Ginn. Boston. 1934.

Kasner, Edward, and James Newman *Mathematics and the Imagination.* Simon and Schuster. New York. 1940.

Rosenberg, Karl, *Das Rätsel der Cheopspyramide.* Band 154. Deutsche Hausbücherei, Österreichischer Bundes Verlag. Wien. 1925.

Smith, David Eugene, *History of Mathematics.* Ginn. Boston. 1925.

Notes on Waldorf Education, Part 1

Notes on the Lectures of
Hermann von Baravalle
Taken by Gisela Thomas O'Neil

"Always let something show out of your sleeve," is the advice given by the master magician of mathematics education in this survey of Waldorf school practice. The notes take us from kindergarten into grade school, and provide a sketch of how Dr. von Baravalle viewed arithmetic and its relation to other subjects of instruction. His lectures went on to treat other topics such as the introduction of the alphabet and geometry; we have followed the lectures in this volume only through his remarks on arithmetic.

The lectures were given sometime between 1952 and 1955. Gisela O'Neil's notes were not a transcript of Dr. von Baravalle's lectures; she recorded only the main points. Her notes have been expanded: Phrases in the notes have been made into sentences for instance. Much of what took place is entrusted to the imagination of the reader. We have found, however, that these lectures contain thoughts that are of vital, contemporary interest.

Waldorf Schools

1. History

John Jacob Astor came to the U.S.A. as a poor, uneducated boy at the age of 16. He sold musical instruments and furs. He also entered the real estate business.

Later he gave up his business and devoted the last 25 years of his life to studies. He became a friend of Washington Irving. In his birthplace Waldorf, near Mannheim, Germany, he donated a school for the instruction of children up to the age of 16. This took place in 1848.

This is the beginning of the *name* "Waldorf school" without connection to Waldorf education.

The main seat of the Waldorf Astoria Company in Germany was Stuttgart. Emil Molt, the director of the company, wanted to make a contribution to reconstruction after the first World War by founding a new school. He wanted to enhance human relations in factories and in daily life through an education which would bring thought-life into action.

He asked Rudolf Steiner, the director of the "College of Spiritual Science" (Goetheanum) in Dornach, Switzerland to start the new school. The founding was on Sept. 7, 1919, with 175 pupils. The Waldorf Astoria factory paid the tuition costs for the workers' children. One of the school's principles was that a child should not be excluded for financial reasons.

By 1938 eight schools had been founded. The Nazis closed the schools because

(i) Waldorf education strengthens and stresses the individual, and

(ii) according to Waldorf school principles there must be international cooperation in cultural matters.

After the Second World War, the Waldorf School was the first cultural institution which was reopened in Stuttgart. In a few years Waldorf education became a leading movement in several countries. There are 65 Waldorf schools in the world; 55 are in Europe:

Germany	25
Holland	7
Switzerland	5
England	9

In Switzerland, the freest European country, 105 public schools adopted the Waldorf Plan. In that country Waldorf education benefits the general education.

[The report of the Association of Waldorf Schools of North America for the year 1993 – 1994 shows 52 Waldorf schools in the United States and Canada alone, and many other schools that are at a stage preceeding actual membership in the Association. It is odd that the lecture notes did not mention any North American Waldorf Schools. Rudolf Steiner School (New York City), Garden City Waldorf School

(Long Island), and the Kimberton School in Pennsylvania surely existed at the time of these lectures. Green Meadow Waldorf School in Spring Valley, New York, might have been too small to attract more than local attention. Dr. von Baravalle often lectured in Spring Valley, however, and one would expect him to taken notice of a Waldorf School there. – Ed.]

2. Why Are There Waldorf Schools?

In founding a new school Rudolf Steiner sought to answer the question: How can we make a school up to date?

The greatest crisis in school starts at the age of 12 when the student loses interest and enthusiasm. What is the reason? Do we teach according to habits from the past? The students are no longer keen for the logical thinking which was taught to their forefathers. They want to touch life, to do something. If we want them to have enthusiasm and interest in their education, we need teachers who have enthusiasm and interest in their subjects.

The faculty of the first Waldorf school was selected according to these principles. The school was to have educators with background and objectivity. Teachers were selected to represent different walks of life.

3. Teachers and Administrators

The Waldorf plan is one in which the best teacher can find his form. It is for teachers who want to make a contribution and who want to make an art out of their teaching.

The teacher can concentrate on his preparation and devote time to it. The students are in the best condition so the teacher is not worn out trying to teach a class that is tired of school. He spends as little time as possible on red tape and is free of unnecessary burdens.

The teacher's life should be as idealistic as possible. The teachers feel that education is a field in which their life can be spent in a worthwhile way.

In Europe the Waldorf school boards consist of trustees who make the school physically possible. A selfless attitude is important. The point of view derived from business should not be carried over into school matters. Leave education its space for creative work.

The Course of the Day and Year
According to the Waldorf School Plan

4. The School Day

The course of the school day reflects the methods that are used in the classroom.

Short periods are used to check homework and to assign new homework. There is little opportunity for the teacher to develop his teaching method, since the short periods do not demand as much skill as the longer periods do. Long periods place an emphasis on good teaching!

In the Waldorf School Plan Rudolf Steiner was concerned with continuity. Nothing is harder than switching subjects, because the second teacher has to undo the first teacher's work and to put another subject as the top priority. This is a loss of energy. Many teachers claim that the first ten minutes in each period are the most strenuous. One subject could be ruined just by an inappropriate sequence of classes, for example, sports and then mathematics. The continuing subject must not be prematurely interrupted, if the students are really expected to concentrate.

The plan intends that a major piece of work should continue to a conclusion and at the same time allows it to end before it becomes worn out. Here is how the school day is planned:

I. *Concentration on the continuing subject.*
(The work is done alone by each student)

II. *Group Activities*: discussion, languages, singing.
(This is a relief from intensive studies, not a repetition of them)

III. *Arts and Crafts*

IV. *Physical Education*
(Done to restore physical balance)

One subject should complete the other. The subjects should not be taught for their own sake but are part of this balance. There should be conditions for the best sequence of subjects during the school day so that one branch of work is not placed at a disadvantage.

The proper plan for a school day is one that corresponds to the natural laws of life.

Mon.	Tue.	Wed.	Thurs.	Fri.
I.	Continuity			
	one of the major subjects			
II.	Distributed			
	foreign language, discussions, singing group activities			
III.	Continuity			
	arts and crafts, science laboratory, geometrical drawing, dramatics			
IV.	Distributed			
	physical activity, recreation			

["Continuity" means that a lesson continues from one day to the next. The term "distributed" describes a lesson that is relatively self-contained. – Ed.]

The main subjects that are in continuity last for 3 – 6 weeks in rotation. This allows them to be complementary.

The advantages of this system are: The students are better able to concentrate on something in depth; there is less tension arising from interruptions; discipline is improved; school life possesses rhythm; there is better contact with the teacher; the energy of the students is economized; intermissions take place naturally; the student is not treated in a mechanical way.

The disadvantages are: Teachers must cooperate with one another; excellent teaching methods are required; teacher selection is difficult; no short cuts are possible; and progress is geometrical.

[After the words "progress is geometrical" which appears among the "disadvantages," the notes showed a small diagram of an exponential curve within coordinate axes. As a matter of fact, when various subjects complement each other properly, the pupils' progress is compounded. This compound, multiplicative rate of progress is what a mathematician means by the word "geometrical" when it is used to refer to a rate of change. It is not clear, however, why this is mentioned as a "disadvantage." There may be an error in the notes. Von Baravalle certainly does not mean to say that students in schools that follow the Waldorf plan learn so rapidly that their teachers must labor to preserve their ignorance.

It has been suggested that these words mean that learning proceeds geometrically, therefore knowledge is acquired much more rapidly in later years than in early ones. This could be construed, from certain perspectives, as a "disadvantage" in the earlier grades. In any event, what is recorded here is what appears in Gisela O'Neil's notebooks. – Ed.]

5. Rotation of Continued Subjects

The continuing subjects rotate in response to the inner reactions of the students; these are subconscious.

The sequence of major subjects alternates between humanities and realities:

    I. *History*

    II. *Mathematics*

    III. *Literature*

    IV. *Natural Science*

I. *History*

The first category, history, treats the human elements, the greatness and shortcomings of mankind, war, and peace.

Students notice a lack of a logical, orderly sequence. So after several weeks of history the continued lesson changes to a subject of a different kind.

II. *Mathematics*

Here there are eternal laws, an orderly sequence of ideas, a well-determined experience of right and wrong, and natural methods of self-correction. Then the students miss the richness of the artistic side of life, so the cycle of continuing subjects moves ahead.

III. *Literature*

In literary studies the artistic side of life appears with the thoughts and feelings of mankind, with happiness and grief.

After a while, the students miss objectivity. They move on to

IV. *Natural Science*

Here there is law and necessity.

6. Plan for a Rotation of Subjects

The rotation of the major academic subjects can be built into the school year, so that the cycle I → II → III → IV → I takes place according to a regular plan.

The following table shows how this can be done in the weeks from the opening of school in September to the Winter vacation that takes place at Christmas.

First term: September – Winter Vacation											
1	2	3	4	5	6	7	8	9	10	11	12
weeks of the school year											
I			II			III			IV		

This pattern may be repeated after the Winter vacation.

Second Term: Christmas Vacation – Spring Vacation											
13	14	15	16	17	18	19	20	21	22	23	24
weeks of the school year											
I			II			III			IV		

Finally, a similar cycle could be carried after the Spring vacation.

Third Term Spring Vacation – Summer											
25	26	27	28	29	30	31	32	33	34	35	36
weeks of the school year											
I			II			III			IV		

7. Subconscious Urges

One tries to teach according to the subconscious urges of the various age levels. The child wants to be guided, wants to feel "school life makes sense." It is wrong for the student to have an attitude that indicates a lack of guidance: Do we have to do whatever we want again?

The teacher must be able to work with a logic that goes to the depths:

He should not need to use a book for the information he requires;

The students are to learn by heart (dates in history for instance);

There should be no waste of time; never start a class with checking a list or looking at homework.

The first minutes in the morning must justify the effort to get to school: The child does not want to miss the beginning of class. The child should feel: yes, it was worthwhile to come in time.

If time is wasted, the child subconsciously feels: Why did I make the sacrifice of getting up and hurrying to school? Here we sit around and get bored. What's good about the school?

The teacher's attention must always be with the group. The children expect something. The first five minutes of the day require the most preparation because of this expectation. The more the teacher is aware of this expectation, the more he will prepare.

The teacher must answer the real question of the child. He should never answer a question which no one has in mind, or neglect a question which anyone does have in mind. For example, concerning grammar: Children in the 7th grade are nearly perfectionists in observation. So in forming grammar lessons you may describe a fact you have observed adding a wish, a hope, or a projection of disappointment. Grammar involves the details that weave in the language as it is really used.

The whole Waldorf plan is an adjustment to the exact response of the various age levels; it fits all countries with slight variations.

Kindergarten

8. Pre-School Level

The younger the child the more different his or her whole reaction is from the adult attitude.

Adult	Words of a Beggar as an Illustration
1) *Thinking* (information)	1) Sir, I haven't eaten all day long.
2) *Feeling* (personal relation)	2) Sir, I feel hungry.
3) *Willing* (action)	3) Will you help me?

This fails completely in approaching a child. Concepts like "good" mean nothing to a child at this age level. The child may ask: "What does a good child look like? Is he or she blond?" How empty it is to search for the meaning of a concept in this way – it will have no meaning.

It is typical for adults to have ideas; conceptions are stepping stones. Thinking appeals to ideals. Life can be a complete "hell" if ideas are missing. Concepts like "good" gain value with age. The older you grow the more concepts and ideals there can be. They provide the "springboard power" for action.

For a child ideals mean so little. It is a weak point in education for teachers to explain at this age level; the whole education is weak if that is the case. What can we do instead?

A child of this age is alert in watching the movements of a craftsman, watching work which is skillfully done. This has tremendous stimulating power. The child sees the craftsman's gesture and absorbs it, plays at being a carpenter with the same swing, the same amount of energy; everything is in the child to minute detail – the gestures remain in the child.

The child does not make an analysis, the action is absorbed; it lives in the child and comes out again. The task of education is to provide opportunities to expose children to these things that they can take in with their whole nature. The skill of years is acquired in one gesture – the child acquires it immediately, spontaneously, without analysis.

The child's approach is not in words; it is through partaking in the gesture.

Willing   The action is watched.

Feeling   Action is assimilated in play, digesting the experience (the younger the child the less imagination).

Thinking   interest

The gift possessed by the child of a capacity for absorption turns education away from explanation to action.

The experience of teaching is like that of the stage: Every gesture has stronger influence than words. The way in which ideals are brought to children is by action that is in harmony with those ideals.

The key fact is that the child absorbs more of what we *are* than from what we *say*.

The child senses if a teacher is tense but talks of peaceful relaxation. The child's awareness leads to his self-education. Real education is self-education. Use clear actions around children. This must be done in the household too, not merely in school.

9. Subconscious State of the Young Child

I do not want to come to this earth as a mere slave. I want to build up from within. These thoughts will appear subconsciously in a child.

Do we want the child to become a free person, looking to life? Do we want it to be able to make decisions later in life? If so, we avoid rules and concepts. At that early age it may be settled whether the child ultimately is an idealist or a conformist. Rules and concepts will bring children into conventions in a way that does not fit them.

Education to freedom is the whole philosophy of life put into gesture – the child will pick it up. Put your philosophy into the surroundings, don't merely say it.

The child's point of view is this: "I do not have to be told: I watch and then I do it."

The major tool of education is spontaneous imitation. Learning a language is done through imitation, from how adults speak; it is not inherited. It is picked up by the child. Adults need explanations; we cannot follow another person's expression with the same intimacy as a child can.

Here is another example: A guest comes to visit. The child watches the phenomenon. Now the guest takes off his coat; his whole philosophy may be expressed in this gesture. "That dignity, that kindness. That is something I haven't seen. How large his hands are. When he talks – that smile – how nice."

That is the approach of the child. The philosophy of the adult is, from the point of view of the child, captured in gestures rather than conveyed by concepts.

10. Work in the Kindergarten

For the ideal balance we have

(I)   activities carried out within the group, and

(II) free play.

(I) Group activities consist of things which give new stimulation, enriching experiences.

(II) Free play involves the mental digestion of the experiences. Not everything should be planned (it is an alarming sign when a child no longer plays alone).

11. Kindergarten Painting

What should be drawn? Not ready-made things, we are too quick to jump ahead. They are neither inspiring nor artistic.

The teacher starts to paint, there's a spontaneous reaction, "I want to paint too!" That urge is in their limbs. Instead of telling, giving explanations, how to use the paint and brush, the teacher acts and the child picks the explanation up actively.

[We presume that the sketch that follows was intended to illustrate "ready-made" picture subjects. – Ed.]

The fundamental elements of painting are the colors themselves

[A colored diagram appeared in the notes with three paint pots, like those above, colored yellow, red, and blue. – Ed.]

We use primary colors and a large brush; the paper must be immaculately clean and white (only the best materials). The senses build themselves. The children experience a liquid flowering of color.

The world is so ready-made for a child that it can experience: "A color is born!" Discoveries come out of themselves; they do not need explanations. From sunlight and darkness chlorophyll develops, as green comes from blue and yellow.

[The notes contained a simple, colored diagram here that we have not reproduced. It was a rectangle with blue at one end, yellow at the other, and graded to green in the center.– Ed.]

This way of painting represents the whole system of development of the plant world. The child has these things in mind without it being nailed down to concepts.

Give half an hour for preparation. The children are absorbed in their preparation. Wait until *everyone* has his partner, brush, and colors. Quietness grows naturally.

[It appears that Baravalle thought of the paint pots as shared by two pupils. – Ed.]

The teacher does not analyze the child's painting, but he leaves them in the state of creative work. Let them find their way into life themselves. Painting brings out their inner conditions. The teacher does not talk about this with the child, he does not make the child an exhibitionist, he does not scold, does not praise, but partakes in and shares in their joy in tender harmony, in an element of balance and harmonization.

12. Reflections on Child Development

In former times the children had more opportunity to watch the movements of craftsmen and to see mother sewing, baking, and mending. Something has been lost with technical development. We have to use finer elements to supplement what has been taken away from the children.

It is a blessing of technical development that the parents may have more time to devote to the children. They can provide finer experiences than could otherwise be done. They could bring the child out into the country, for instance. They should put the right emphasis on what counts for the child: to see people in the scene of rural life.

There should not be an examination afterward – it would bring a double disappointment.

The children are very observant – but we should not force memories on them. Let them observe. Their observation is better than ours. Don't ask them for definitions. Never bring in anything too early on account of convention or ambition, wait until the child is fully ready.

To bring in something too early has a doubly negative aspect:

1. disappointment – it does not help in the moment;

2. we spoil it for the child who looks forward to experiencing it at the proper time.

We can make positive use of the child's expectation for what is to come. At the school year's beginning each teacher gives a few indications of what the students are going to have. A child in the lower grades develops the attitude: "Oh, when I come to the 6th grade, then I shall have geometry with all those beautiful drawings."

The curriculum of the Waldorf school has been carefully planned so the children are just awakened to their subjects. What a blessed place: When I feel an urge for something, it will be done.

In this golden age of childhood, we do not introduce prizes or money. We let the grinding money world pass by so that the basis for idealism can be lain in this age. Good will often produces terrible results. An American attitude is: I want my child to be happy, it shall learn the worth of the dollar and take an early root into this world of money. This attitude is likely to produce the opposite result

A European attitude is: My child has to face a hard life. I must harden these children in order to prepare them for life; their life shall be as rough as my life was for me. This attitude may affect their health.

[These lectures took place just after Europe had suffered from great wars and economic crises in the preceeding decades. – Ed.]

If we observe life, it will often be striking; but to do so we must open ourselves for observation.

Disappointments are easily projected: a wish, a hope, having a different child in mind than it really is, treating a child as who you want him to be. Take the child as he is. Do not intellectualize your approach to children. Instead of projecting concepts onto them, let them have their own world.

If parents would have this attitude, then children would even start to breathe differently in many families.

This time is free and reserved for childhood. If the children quarrel, then the teacher realizes that there is too little flow to the play; the children need stimulus.

The teacher has open organs for everything that is going on – but the child does not realize it. The teacher should be active himself so the child can imitate him. Don't push. Don't command.

Being naughty is the expression of being under defense. Naughty children may have too little opportunity to be themselves.

13. Kindergarten Facilities

The kindergarten is a prolonged playground. Use all the facilities which nature can give: hills, bushes, old trees, tree trunks.

Hills give the opportunity to run up and down. Have a spiral path for going up. Paths should be in curves so they can play: good-bye, come again, etc.

The modern point of view is: use iron, concrete, asphalt, strong wagons – this way the children can't damage anything. We want to indicate finer things. We make the whole place inspire delicacy; the children learn to appreciate it, they take care of it. We reckon with the finer qualities: friendliness, a light green color, flower beds, inspirational things. Let the eyes of the children explore and find the natural situations of play. Plant trees close together so they can slip in and out. They can develop their balance on old tree trunks.

A play house is unfinished and has open windows and doors, so the children can slip in and out. Sand boxes often are enclosed on the bottom; then they must be covered for protection from rain. The whole thing becomes a breeding place for bacteria. Have a wooden frame but without an enclosed bottom so that it drains directly into the soil. Water passes through the sand and disappears. The sand is cleaned by rain, sterilized by sunshine. Such a sand box is natural.

The toys are unfinished. We should leave the children some flexibility. A doll that is too finished has one fixed expression; since it shares life with the child, it needs to change its expression. Indicate something and then leave it unfinished.

The child finishes it with his imagination. The simpler the toys, the more the child contributes to the development of the imagination.

14. The Early Years Form a Basis

The form of adult humor in which an animal is used as a caricature of human weakness (in movies, television, and the funnies) may look funny and please us, but it only subtracts from the values of the child – it only brings confusion. The animals really represent a realm of nature that is on a different level from the human one.

In plastic work we use clay or wax as the material and do no naturalistic modeling. The child puts his finger in and sees how the mass grows. He experiences the abilities of forms to change. He wants to be creative.

Eurythmy, excursions (for watching), and storytelling are suitable kindergarten activities.

All instruction comes out of creation. This is the source, arising in this time of life, for the ability to take initiative later. When activity, will, is lacking later in life, we may trace it back to those years.

The ability to take initiative is closely related to free individuality. Life has devoted this first period to it, before the intellect comes. If we want the children to show initiative, to become contributors to life, then let us leave them to form their lives from within. Further stages will take place. One stage closes: a new one opens. The growth of a child is not the progress of addition. Qualities come and withdraw. There are gifts which come and go (for example, the wonderful gift to learn a foreign language with ease at the ages of 7 – 9). The gifts of imitation, of following, of partaking directly in the surroundings, of direct contact with action, this is limited to the period before the change of teeth.

The change of teeth is the most obvious time in which a change that takes place throughout the organism is also accompanied by an inner change.

The new teeth are important tools until the last years of life. The harder teeth are ready to meet physical demands.

Spontaneous imitation goes – the child develops a different personality.

It is a tendency in our time to focus too much on what a child learns – in a time when the child is primarily building up his body. Steiner wrote: "Before the change of teeth in the seventh year, the human body has a work to perform upon itself which is essentially different from its tasks in all other periods of life – namely, to mold the physical organs into definite shapes. Growth continues of course in the later periods of life, but it is then based on the forms which were developed in the first life period. If the right forms have been developed, right forms will grow, if misshapen forms have been developed, misshapen forms will grow. We can never repair what we have neglected to do as educators during the child's first six or seven years."

Eurythmy is taught to preserve the lightness of the limbs. Some people show that the elasticity of life remains with them (it provides an inner balance); their every step shows it. They do not drag their bodies like a weight. To lose vitality is due to the handling of the first years. We must restore this vitality. The limbs must be instruments for the use of the will.

Imitation, growing into life, seeing people do what they themselves will someday do, is the free way of development for a child. We should hold our explanations until the child is keen for thoughts through words. Preserve the intellectual answers until later and the child will meet them with enthusiasm.

Joy is a constructive building principle; we should not display any sorrows before the children. We should show our love for the children in an honest, unaffected way. The love must be really sincere; children dislike all conventionality in love. If we have to meet sorrows or grief, we should not try to become different people and behave gaily – but try from the wealth we have to give them more. We are individuals – we have potential authority without intellectualization or concepts. The way we present this authority is through the sincerity, the reality, the truth that we carry into life.

Concerning religion, many parents are anxious to present to their children that for which they have devoted their lives. What is helpful for the child is only the attitude of devotion. Reasoning about it hurts and very often has tragic results.

The general approach, to storytelling, is action, will, doing. Every sentence should carry the action farther. You can correct wrong action by right description.

Don't give lessons on transportation! Very few people grow up nowadays without knowledge of cars, subways, etc. The child is much more interested in finding his position in the world. The story has simple action, first in one direction then another direction. This is enough. Later on, color and image are needed to reach their hearts, filling their hearts and minds.

At the age of 12 or 14, and later on, some children are intellectually tired, their interest is down almost to "zero." We have to see this without any illusion. Something that brought interest to a classroom ten years ago leaves them cold. We have to reform our subjects.

When intellectual joy and enthusiasm is torn down, it is the fault of former education which did not bring them to a state where they appreciate great knowledge. The approach to the child should suit the age:

age 14 – 21	thinking
7 – 14	feeling
0 – 7	willing.

Grade School

15. The Second Period of Life

In grade school, the second period, educational goals are approached through feeling. Feeling asks:

How did you do it?

How did you present it?

Teaching goes completely through the heart. To reach the heart is more difficult than to give a source of knowledge. This is the realm in which teaching is an art.

The attitude of a college student is: My professor has a lot of background even if it is clumsily presented.

A child of grade school age depends completely on the relationship to people. If the child is reached in the heart, he will "go through fire" for the teacher. But if that is missing, there will be constant trouble. The child has a natural wish to admire; and it will become a healthy, imaginative, lively personality afterward when it gets guidance through the teacher's art of presenting life.

Two key words: "image"
                 "rhythm"

16. The First Day of School

The first day of school is a unique moment in the child's life. The child comes out of the hands of parents and other authorities arise.

We bring this in the most artistic and tactful way: "You are growing from a smaller family into a bigger family." Each one feels included. Each feels the care that is there.

Why do you come? Why do you go to school? You come so that you will be able to do what your father, your mother, and the grown-ups do. They can read the paper and cook. They can make windows, buildings, and chairs – for you to sit on. You would get tired without a chair. You come to learn useful things.

To make a social appeal, the tables are shiny and the floors clean. The central idea is that the child is in the midst of society, growing from a smaller family into a bigger community and many people contribute to the things that appear in your life.

The children are all wondering what will happen. To answer, provide a digest of what they see. Tell them what is in the background. The contributions of millions of people are in such a school. It is important to stimulate the undertone of feeling so that they will like to work.

For example, we are using white paper in a certain class: It was a long procedure to make it. Several people touched it. Yet it is clean! Shall we be the first ones who make it dirty?

Appeal to their enthusiasm for the world of grown-ups.

17. Arithmetic

In teaching the multiplication table, the "old-fashioned" way was: Good drill is necessary. The modern way is: no drill at all. We accept neither of the two!

When counting the child experiences the order of necessity arising from a sequence.

$$1 \quad 2 \quad 3 \quad 4 \quad 5 \quad \cdots.$$

If I do not keep this order, if I say, "1   2   5   4," everybody will feel that I am wrong in counting this way.

We can produce an ordered sequence in threes:

$$1 \; 2 \; \mathbf{3} \quad 4 \; 5 \; \mathbf{6} \quad 7 \; 8 \; \mathbf{9}$$

by jumping or clapping on the emphasized numbers. It feels as though music and play are in it. The children must feel it. Feeling is their vitality. There is rhythm in the variation; they like it. They don't feel dull, but they feel relief.

The arithmetic must be rooted in their system. The world of music is reflected in a fourfold rhythm:

$$1 \quad 2 \quad 3 \quad \mathbf{4}$$

Why does the clock go up to twelve?

A wrong method is to say: Twelve is a special number. It was used in very old measurement, up until the 13th century. It can be divided by 2, 3, 4, 6, and so on.

There must be a rhythmic experience of this: 12 is the meeting point for 2, 3, 4, and 6, the four numbers which, of all the numbers in the world, we meet first (the next meeting point is 60).

The child feels: "Yes, it is so. I have been there where that was proved." The rhythmic weaving has become part of life itself.

Order is fundamental in mathematics.

$$1 \quad 2 \quad \mathbf{3} \quad 4 \quad 5 \quad \mathbf{6}$$

[The meaning of the notes is obscure here. In reviewing the idea that arithmetic should be introduced by rhythmic counting, one and two seem to have been called "closed" and three "open." The remark "blow them away" appears. This was probably intended to encourage a dramatic emphasis of the rhythm. Dr. von Baravalle mentioned eurythmy gestures for enclosure and opening in connection with rhythmic counting. He probably meant to recommend the use of these gestures during arithmetic lessons. – Ed.]

Children want to explore. The teacher may let them find the scale of 7, for example.

1 2 3 4 5 6 **7** 8 9 10 11 12 13 **14**.

In second grade the situation is such that the teacher will want to repeat the lessons. If a lesson was really good, it can be destroyed by repetition in the same way. You must find a new way of doing it.

18. The Fundamental Block of Arithmetic

1	2	3	4	5	6	7	8	9	10
2	4	6	8	10	12	14	16	18	20
3	6	9	12	15	18	21	24	27	30
4	8	12	16	20	24	28	32	36	40
5	10	15	20	25	30	35	40	45	50
6	12	18	24	30	36	42	48	54	60
7	14	21	28	35	42	49	56	63	70
8	16	24	32	40	48	56	64	72	80
9	18	27	36	45	54	63	72	81	90
10	20	30	40	50	60	70	80	90	100

It is a challenge just to write it all down neatly. This is done for pictorial memory. Adults often overlook the effects of repeated action.

"Could we do this again?"

When it is repeated, we must do it with something new and stimulating, so that it does not get stale. Some bright minds are active while writing. They can discover some of the facts about the symmetries of the table.

For example, compare the final digits in the 9's scale with the 1's scale: they run backwards. These digits are listed in adjacent columns in the table that follows. We may compare the numbers that are adjacent as shown by the arrows.

The 2's scale repeats itself twice. This repetition is shown in the table of last digits as well. The final digits of the 8's scale go backwards when compared with the 2's.

There will be discoveries from the first day on.

You must be an explorer.

FINAL DIGITS

9 scale	1 scale		2 scale	8 scale
9	1		2	8
8	2		4	6
7	3		6	4
6	4		8	2
5	⇐⇒ 5		0 ⇐⇒	0
4	6		2	8
3	7		4	6
2	8		6	4
1	9		8	2
			0	0

3 scale	7 scale		4 scale	6 scale
3	7		4	6
6	4		8	2
9	1		2	8
2	8		6	4
5	5		0	0
8	2		4	6
1	9		8	2
4	6		2	8
7	3		6	4

Child : "I don't like those numbers they don't have enough order."

Rhythm arises when a regularity goes forwards and backwards.

1	2	3	4
2	4	6	8
3	6	9	12
4	8	12	16

$$1 + 2 + \underline{3} + 2 + 1 = 9$$

$$1 + 2 + 3 + \underline{4} + 3 + 2 + 1 = 16$$

$$1 + 2 + 3 + 4 + \underline{5} + 4 + 3 + 2 + 1 = 25$$

Mathematics reveals to the student that *the world is ordered.* The mind is grasping the whole universe. We are in a world of order. Be assured, you are not alone; you are part of something.

Students detect the wonders of the universe with a *spirit of exploration.* They grow into the world of wonder and amazement; they feel the whole. "I want to learn to be a part of that order."

They feel that there is giving and receiving. *When I am an adult, I don't want to be a receiver. I want to prepare myself to make the circle complete.*

Teacher: give the material, the indications, a little remark – they will do the rest. Do not say: "It's wonderful."

19. The Use of History

The stimulating part of teaching is the subject itself. Let the subject talk.

The teachers task is to find out: Where is the corner in the subject from which I can start to introduce it?

History has introduced it first. The content has to be recreated.

20. Albrecht Dürer and the Magic Square

Albrecht Dürer stood at the time when the modern age began. He was called a "universal genius," like Leonardo da Vinci. They were both able to extend their work to both science and art. Dürer was a scientist, artist, engineer, and one of the inventors of linear perspective. One of his most famous pictures is the *Melancholia*. If we call it "Melancholia," using the conventional title, we give it a completely wrong interpretation. Dürer wrote the Latin "I" after it. "I" is the imperative form of the Latin word "ire," to go. "I" means "go!"

[The figure reproduced here shows the bat-like creature taken from the dawn sky of Dürer's woodcut that carries the banner meaning *melancholia go!* – Ed.]

It is no reflection of the melancholic mood; it expresses the flight from melancholy. The dark ages are coming to a new life through invention! In science and art new inventions break in and the melancholia which darkened the human mind like a cloud had to flee away. The new art of the mind began!

In this picture "Melancholia go" all sciences are presented – no one is left out – the field of mathematics is present too. The whole picture is conceived as a new light coming in and opening the world.

The magic square shown below appears in this same picture of Dürer.

16	3	2	13
5	10	11	8
9	6	7	12
4	15	14	1

Arithmetical relations are in this picture. But your mind must be active, you must make an effort to get them. That is characteristic of the modern age.

The solution of the magic square is found by combining the numbers in the right way; the result is always 34.

There is hardly any practical application of the magic square – it lacks a practical side. It is from the world of pure numbers in which the Greeks lived. The activity of the mind is in directing the numbers. The mind and the world are both mathematical. In our time there is an overemphasis on so-called applications. This reverses the proper order of things.

Concern with practical applications is the weakest point in our "modern" way of teaching. We imagine that we are stimulating the children; meanwhile the child's mind goes to the pure operation. The child's mind works differently from the way that grown-ups think.

The proper method: No "practical applications."

Let the child start from within, from his own world.
Let him find: The world is so.
Let the field be his own.

Allow the children to live in a world of pure numbers.

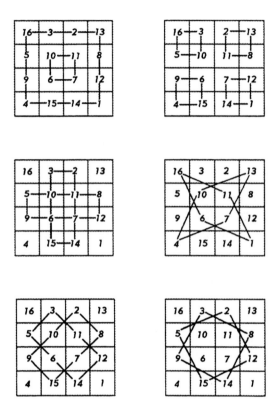

Here is the way in which Dürer constructed his magic square.

16	15	14	13
12	11	10	9
8	7	6	5
4	3	2	1

Begin with the numbers from one to sixteen placed in a four by four array in descending order. Leave the four corner numbers protected where they are, then turn to the six positions that are along an outside edge but not at a corner. These are to be interchanged with the numbers opposite them. The following tables show the edge numbers in dark boxes. In the table on the right the diametrically opposite pairs have been interchanged.

16	15	14	13
12	11	10	9
8	7	6	5
4	3	2	1

16	2	3	13
5	11	10	8
9	7	6	12
4	14	15	1

Finally, the two central columns must be interchanged. This gives the magic square.

21. Another Important Historical Event

In 1636 analytic geometry was invented by Descartes showing the close tie between algebra and geometry, between the order in numbers and order in space.

22. Order in Arithmetic

The world is an orderly whole. We have to combine what we sense with our thoughts in order to understand; we have to step actively in. The little ones do it too.

order in space = *image*

order in time = *rhythm*

Image and rhythm are the two key words for the second period of life.

Little children want to grasp the moon with their hands. Only gradually do they develop perspective. "I can do it!", the child says, with a joyful mind. They are most interested in what directly concerns themselves.

They will find order in arithmetic.

The following computations show how computations can be arranged to provide order in the very arrangement.

$$1 \times 9 + 2 \quad = \quad 11$$

$$12 \times 9 + 3 \quad = \quad 111$$

$$123 \times 9 + 4 \quad = \quad 1{,}111$$

$$1234 \times 9 + 5 \quad = \quad 11{,}111$$

$$12345 \times 9 + 6 = 111{,}111$$

$$1 \times 8 + 1 = 9$$

$$12 \times 8 + 2 = 98$$

$$123 \times 8 + 3 = 987$$

$$1234 \times 8 + 4 = 9876$$

$$12345 \times 8 + 5 = 98765$$

$$123456 \times 8 + 6 = 987654$$

23. Your Favorite Number

Which number do you like?

Suppose the answer to be 7.

$$
\begin{array}{r}
12345679 \times 63 \\
\hline
74074074 \\
37037037 \\
\hline
777777777
\end{array}
$$

[In this example, as throughout his lecture notes, Dr. von Baravalle carries out handwritten multiplications in a style that is different from that taught in American schools. We have left this computation and the one that follows in their original form.–Ed.]

Suppose the favored number were 2.

$$\underline{12345679 \times 18}$$
$$\underline{98765432}$$
$$222222222$$

The sum of the digits of the number 111,111,111 is divisible by 9; and 222,222,222 is its double. When 111,111,111 is divided by 9 the result is

$$12345679.$$

[The notes contain a long division here along with some marks which point out that one expects to find, in the quotient above, an 8 that is missing. The reader should carry out the division in order to see that it does have an interesting rhythm and a missing 8. If your favorite number is 6 you multiply 12345679 by 6·9 = 54 – Ed.]

Their interest will be attracted to such a problem.

24. The Magician's Advice

The teacher should always "let something show out of his sleeves".

[We understand this remark to compare an arithmetic teacher to a stage magician who reveals the secrets of his trade, while retaining an air of magic that will stimulate the mind toward further explorations – Ed.]

25. Surprises

251	958	896
324	745	784
417	836	983
119	794	771
1111	3333	3434

Select the last number.
Each time there is a surprise! It is not to bring advantages to the teacher, but to keep the interest of the students.

Child: "How do you do that?"
"I'll prepare for the next class."

We don't let it get dull. We bring them into the picture.

Mentally the child is to think: "I want to do it."
"I can do it."

[The sums given at the beginning of the section are problems given
to pupils. The teacher chooses the last number, so that patterns
appear in the answers. But the patterns do not become stale; they
change into broader patterns. The words "advantage to the teacher"
probably refer to the fact that corrections are much easier when
patterns of this kind appear. But that is not why it is done. The
arithmetic lesson reflects the larger world of concepts by offering
interesting patterns to the mind. – Ed.]

26. The World of Numbers

We let arithmetic be something from a world of its own.
Let the subject speak, then it is understood. The key words for
presenting arithmetic are: rhythm and image. In this way it is
more true, more familiar, and more closely related to the
attitudes of former cultures.

"Image" resembles the attitude of the ancient Greeks in
which science was closely related to art. There was a feeling
for the beauty in things.

Every exercise should be written neatly and spaciously. "I
can't stand sloppiness", is a good attitude for study.

27. Addition

Addition starts with counting. The child should feel: "I can
go on for so long!"

Then, we can start counting from 12 or 25. Next, limit the
counting to a certain number of steps. Go five steps forward,
for instance.

12 13 14 15 16 17

The last step is to transfer the gesture to symbols. Symbols are condensed, miniature gestures. We can count on and on and on, then stop and underline the last one.

The use of the fingers amounts to a coordination of the mind and the limbs. Rhythm is something both mental and physical.

28. Subtraction

First, we can ask the children to count backwards: 20, 19,⋯, 1. Start from somewhere in the middle, 18 say, and begin there.

To introduce rhythm, we can go backwards a certain number of steps, five for instance.

$$15 \quad 14 \quad 13 \quad 12 \quad 11 \quad 10$$

The child's attitude, "I can do it!" is more important than any practical applications. We let even the humble, little minus sign evolve.

The method is:

first we know	then, we write it down
first it is dynamic	then, it is static
it starts in gesture	then, it becomes a rule
first it is an activity	then, it is a concept

$$3 + 4 - 2 + 5 = 10$$

Start at three, go four steps forward, two steps back, and then five steps forward. Ten is the result.

Motion involves a discipline. If you do it with steps, right away you feel a world of order. The dynamic gesture leads the static presentation.

29. The Pacing of Lessons

$$34 \qquad = 30 + 4$$
$$28 \qquad = 20 + 8$$
$$15 \qquad = 10 + 5$$
$$\underline{12} \qquad = \underline{10 + 2}$$
$$\qquad \qquad 70 + 19 = 89$$

Couldn't we do that in a shorter way?

$$34$$
$$28$$
$$15$$
$$\underline{12}$$
$$19$$
$$\underline{7} \qquad \text{The seven is polite, it waited for the 19.}$$
$$89$$

Let the process evolve gradually. Show them gradually. Don't show the last step first!

30. Multiplication and Division

The following table presents some facts concerning the number 6.

number	Is it in the six table?	six table	how much too big?
40	no	36	4
26	no	24	2
57	no	54	3
14	no	12	2
25	no	24	1
18	yes	18	0

These facts of multiplication are used as component parts of the division that follows.

$$\begin{array}{r} 243 \\ 6\overline{)1458} \\ \underline{12} \\ 25 \\ \underline{24} \\ 18 \end{array}$$

Make long divisions truly long.

$$6\overline{)1116452734284}$$

The longer, the better. Let paper go under the board and across the floor.

"I can do it," is the motto of arithmetic. Then there will be no inferiority complex.

For slower learners:
Don't rush them. They will get it. Look at the multiplication table until it becomes secure. This will help keep the image in memory. They will use it until the moment comes: "I can do it without looking."

31. In the Classroom

Learning is a conquest of the mind: We challenge – they conquer.

Arithmetic must be full of vitality. We must stop if it gets dry. Rudolf Steiner said: I would consider it a bad arithmetic lesson if there were no laugh in it.

A class needs to be conducted like an orchestra. There is a need for continuity throughout a period of time (a unit). Some things cannot be left to the next day; it would be like stirring something up and then not finishing it.

When a new unit is begun: We do not repeat but present what is new. The spirit is: "We climb on, we conquer." The child should think: "I never loved it so much."

32. Notebooks

Homework is the continuation of what was done in class. The purpose of notebooks is

    1) for working in class, and
    2) to preserve the main contents of the lessons.

The child asks: "Which one shall we keep in our clean book?"
["Clean book" is a literal translation of the German "Reinschrift" which refers to the good copy, the copy that is to be kept. – Ed.]

Headlines come last.

In the Bible things were named at the end. The activity is first; the names are last.
    The steps are
        dynamic activities;
        condensed activity;
        symbols (what we did); and
        terms (names).

33. Feelings

There should be a constant feeling of joy, of achievement. Without that feeling the subject becomes dry. The teacher's warmth of heart is connected with his work.

If we are able to take iniative, that is something that goes back to our kindergarten age. How we as human beings feel connected with work, do we feel connected with it or not, that goes back to the elementary grades. To grow enthusiastically, that is something that we may never lose.

34. Rates of Learning

The different speeds of the pupils must be handled methodically by the teacher. The teacher has the attitude:

This is what is brought to me as an educator. I don't want to try to make the world any better than God created it.

It is best to have some combination of fast and slow pupils. One is not complete without the other. A social matter is involved: Some people can give help, some people need help.

A child carefully watches how a teacher handles this problem. A teacher can take the attitude: The whole class celebrates when a slow learner is able to do a task.

The faster pupil is challenged to do more. The slower pupil is encouraged to do a lesser task. This is the most natural approach.

One of the lowest instincts in society is jealousy! The teacher has to be aware when these instincts come up. By creating the right atmosphere the pupils will not be jealous, but they will feel respect for the one who is able to do better than the others. It is a big struggle in life to achieve something – that is a victory. "We have a boy in the class who can do everything."

The Biblical story of Joseph and his brothers can be used against jealousy.

Bibliography of Hermann von Baravalle

Many of von Baravalle's publications appeared in small editions, from obscure publishers devoted to Waldorf school pedagogy, and sometimes in two languages. While this bibliography can safely be assumed to include all of the major titles, and is surely superior to previous ones, some bibliographic details are not certain. For example, the publications are in chronological order, though it may be that there has been an earlier edition that has escaped our attention. It has been necessary to group some items out of chronological order at the end because of uncertainty about the date of publication.

[1921a] *Zur Pädagogik der Physik und Mathematik.* Der Kommende Tag AG. Stuttgart. [There was an edition from Waldorfschule-Spielzeug und Verlag. Stuttgart, 1928]

[1921b] Raum und Zeit. *Die Drei.* 1. p. 352–355..

[1924] Einige Gesichtspunkte für den ersten Unterricht in der Buchstabenrechnung. *Die Freie Waldorfschule Mitteilungsblatt.* 4/5. p. 51–54.

[1925] *Der Unterricht im Rechnen und der Geometrie als Erziehungsmittel zu innerer Freiheit.* Waldorf-Spielzeug und Verlag. Stuttgart. [17 page booklet. Number 1 in a series: *Aus der Pädagogik der freien Waldorfschule: eine Schriftenreihe.*]

[1926a] *Geometrie in Bildern: Mappe I, Bilder aus verschiedenen Gebieten der Geometrie.* Hermann von Baravalle. Stuttgart.

[1926b] *Geometrie in Bildern.* Hermann von Baravalle. Stuttgart. [This is a 12" by 16" folder containing unbound sheets. There are several different topics: Pythagoreische Lehrsatz, Vierecke mit zugeordnet Dreieck, etc. An edition from 1927 carries the title *Mappe II, Pythagoreischer Lehrsatz mit beweglichem Modell.*]

[1926c] *Geometrie in Bildern: Geometrie des Dreiecks.* Hermann von Baravalle. Stuttgart. [The format of this publication is like that of 1926b. The 1927 edition says *Mappe III, Geometrie des Dreieck.*]

[1928a]   *Geometrie und Körperbewegung: Ein Beitrag zum Aufbau eines gesunden Verhältnisses von körperlicher und geistiger Erziehung.* Waldorfschule-Spielzeug und Verlag. Stuttgart. [22 page booklet. Number 3 in a series: *Aus der Pädagogik der freien Waldorfschule: eine Schriftenreihe.* J. Beck lists a slightly different title in a catalog in 1930, *Geometrie und Körperbewegung: Der Unterricht im Rechnen und der Geometrie als Erziehungsmittel zu inner Freiheit.* We have not seen a copy and do not know if it differs sufficiently from [1928a] to deserve a separate entry.]

[1928b]   Goethes methodische Prinzipien von Urbild und Gegensatz im mathematischen Unterricht der Oberklassen. *Zur Pädagogik Rudolf Steiners.* 2. p.78–85.

[1930a]   *Der Sternenhimmel über und unter uns.* Verlag J. Beck. Stuttgart-West. [An edition is listed in the Verlag Emil Weises Buchhandlung catalog of 1940. Third edition: Verlag Freies Geistesleben. Stuttgart, 1958; and Troxler Verlag. Bern, 1958. 19pp.]

[1930b]   *Einführung in die Erscheinungen am Sternenhimmel.* Verlag J. Beck. Stuttgart-West.

[1932a]   *Zahlen Für Jedermann.* Franckh'sche Verlagshandlung, W. Keller. Stuttgart. [There was a new edition in 1939. A revised edition appeared in 1949 and 1959.]

[1932b]   Zum Lehrplan für darstellende Geometrie und geometrisches Zeichnen. *Erziehungskunst.* 6. p.51–61.

[1934]   Zusammenwirken von Innen und Aussen im menschlichen Erkennen und seiner Ausbildung. *Menschenbildung. Erziehung und Unterricht vor den Aufgaben der Zeit. Bewusstseinsfragen der Padagogik Rudolf Steiners.* Basel. p. 279-287.

[1935a]   *Das Reich Geometrischer Formen.* Verlag Freie Waldorfschule. Stuttgart. [There seems to have been an edition from Verlag Emil Weises Buchhandlung, Dresden.]

[1935b]   *Kalendar 1936 der mathem.-astronomischen Sektion am Goetheanum.* Math.-Astron. Sektion. Dornach. [Von Baravalle was at this time the official leader of the Math.-Astron. Sektion of the Goetheanum even though he was in America.

After 1940, however, he could no longer rely on the mail so these star calanders were done by Louis Locher.]

[1936] *Kalender 1937 der mathem.-astronomischen Sektion am Goetheanum.* Math.-Astron. Sektion. Dornach.

[1937a] *Kalendar 1938 der mathem.-astronomischen Sektion am Goetheanum.* Math.-Astron. Sektion. Dornach.

[1937b] *Die Erscheinungen am Sternenhimmel.* Verlag Emil Weises Buchhandlungen. Dresden. [The first edition is [1930b]. This second edition, from Novalis Verlag, was expanded. The copies from Novalis that we have seen lack dates. Third ed.: Verlag Freies Geistesleben; Stuttgart; 1958. Fourth ed. 1962.]

[1938] *Astronomische-naturwissenschaftliche Beiträge für das Jahr 1939.* [We have not seen a copy. It was listed as in the catalog of Emil Weiss, Dresden, 1940, but was probably published in Switzerland.]

[1939a] *Die astronomischen Erscheinungen des Jahres 1940.* [The note to [1938] applies here too.]

[1939b] *Physik 1. Buch: Mechanik der Bewegungen, der Kräfte und der festen Körper, Flüssigkeiten und Gase.* Verlag Emil Weises Buchhandlung. Dresden.

[1940a] *Physik 2. Buch: Physik der Wärme und Kälte, Magnetismus und Elektrizität.* Verlag Emil Weises Buchhandlung. Dresden. [It is astonishing to see such a publication in Germany at this time, since the National Socialist authorities generally supressed everything associated with Rudolf Steiner. The Emil Weises firm of Karl Eymann is said to have been protected by Rudolf Hess. It probably closed shortly after publishing this book.]

[1940b] Mathematiche Denkarbeit. *Mathematische-Astron. Blätter.* Dornach, Switzerland. vol. 2. p. 44–45.

[1940c] *The Astronomical Phenomena of the year 1941.* Hermann von Baravalle.

[1941] *The Astronomical Phenomena of the year 1942.* Hermann von Baravalle.

[1942] *Introduction to the Astronomical Phenomena with Astronomical Almanac 1943.* Hermann von Baravalle.

[1944] Psychological Points of View on the Teaching of Arithmetic. *Math. Teacher.* 37. p. 341–346.

[1945a] Geometric Drawing. *Eighteenth Yearbook of the National Council of Teachers of Mathematics.* Teacher's College, Columbia Univ. New York.

[1945b] The Number *e* – the Base of Natural Logarithms. *Math Teacher* 38. p. 350–355.

[1946a] Demonstration of Conic Sections and Skew Curves with String Models. *Math Teacher.* 34, 1946. p.284–287.

[1946b] Present Developments in the Teaching of Mathematics. *Elemente der Math.* 1. p. 17–20.

[1946c] *Physik.* D.C. Heath. Boston. [This is a 51 page booklet for use in teaching German to technical students. The editor, Siegfried Muller, provided vocabulary notes.]

[1946d] Continuous Transformation of Regular Solids. *Math Teacher,* 34. p. 147–154.

[1946e], [1947a], [1948a] Dynamic Beauty of Geometric Forms. *Scripta Math.* 12. 1946. p. 294–297, *Scripta Math.* 13. 1947. p. 154, 235, and *Scripta Math.* 14. 1948, p.72.

[1947a], see [1946e].

[1947b] Centroids, *Math Teacher.* p. 241–249.

[1948a], see [1946e].

[1948b] The Geometry of the Pentagon and the Golden Section. *Math Teacher,* 41, p. 22–31.

[1948c] *Geometry at the Junior High School Grades and the Waldorf School Plan.* The Waldorf School, Adelphi College. Garden City. New York.

[1948d] Transformation of Curves by Inversion I. *Scripta Mathematica,* 14, p. 113–125,and p. 267–272.

[1949] *Zahlen Für Jedermann.* Franckh'sche Verlag, W. Keller. Stuttgart. 2nd ed. [Revised from 1932a]

[1950] *The Teaching of Arithmetic and the Waldorf School Plan.* Waldorf School Fund, New York. [There was another edition in 1950 from the Waldorf School of Adelphi College. Garden City. Third edition: Waldorf School Monographs. Englewood, N.J., 1967. Also Rudolf Steiner College Publications. Sacramento, Calif. 1991]

[1951] *Physik als Reine Phänomenologie in Drei Büchern, Drittes Buch: Akustik und Optik.* Troxler-Verlag. Bern.

[1952a] The Number π. *Math Teacher.*55, p. 340 – 348. [Reprinted in this volume.]

[1952b] *Methodische Gesichtspunkte für den Aufbau des Rechenunterrichts in der Volksschule.* Verlag Freies Geistesleben. Stuttgart. [This closely resembles [1944].]

[1952c] *Perspektive.* Verlag Freies Geistesleben. Stuttgart.

[1952d] *Rudolf Steiner as Educator.* [This publication from Adelphi College may have had an earlier edition. We have not yet seen a copy.]

[1952e] Die Geometrie der Schattenbewegungen, ein Beitrag zum Astronomieunterricht. *Die Menschenschule.* 26. 1952. p.142–148.

[1953a] *Physik als Reine Phänomenologie, Erstes Buch. Mechanik.* Troxler-Verlag. Bern. 1953 [This seems to be [1939b] with a new title. In 1993 there appeared: *Physik als Reine Phänomenologie, Band I. Mechanik, Wärme und Kälte.* Verlag Freies Geistesleben, Sttutgart. It was advertised as an "aktualisierte Neuausgabe."]

[1953b] *Physik als reine Phänomenologie, Zweites Buch. Wärme und Kälte.* Troxler-Verlag. Bern. [This seems to be an edition of [1940a] with a new title. In 1993 there appeared: *Physik als Reine Phänomenologie, Band I. Mechanik, Wärme und Kälte.* Verlag Freies Geistesleben, Sttutgart.]

[1953c] The Geometry of the Pentagon and the Golden Section. *Science Journal, Adelphi College*, 1953. p. 5–11. [This is an abbreviated version of 1948b]

[1954]   A Regular 24-Sided Polygon with all its Diagonals., *Newsletter Association of. Math Teachers, NY State.* IV. 1. p. 1–4.

[1957a]   *Geometrie als Sprache der Formen.* Novalis Verlag Freiburg. [The book has 411 figures. The 1963 and 1980 editions were from Verlag Freies Geistesleben, Stuttgart.]

[1957b]   *Rechen-Unterricht und der Waldorfschul-Plan.* J. Ch. Mellinger-Verlag, Stuttgart. [This is an enlarged edition of *The Teaching of Arithmetic and the Waldorf School Plan* from 1950.]

[1958a]   Der Unterricht im Rechnen und in den mathematischen Fächern an den Waldorfschulen. *Der mathematische Unterricht für die sechs- bis fünfzehnjährige Jungend in der Bundesrepublik Deutschland.* Fr. Drenckhahn (ed.). Vandenhoeck & Ruprecht, Göttingen.

[1959a]   *Introduction to Physics in the Sixth Grade of the Waldorf Schools: The Balance between Art and Science.* Waldorf Schools Fund. New York. [Second ed. 1967]

[1959b]   *Geometric Drawing and the Waldorf School Plan.* Waldorf Schools Fund. New York. [2nd ed. enlarged: Waldorf School Monographs. Englewood, N.J.,1966. Also Rudolf Steiner College Publications. Sacramento, Calif.. 1991]

[1959c]   *Introduction to Physics in the Waldorf Schools. The Balance between Art and Science.* Waldorf Schools Fund. New York. [Waldorf School Monographs. Englewood, N.J., 1967. 2nd ed. (3rd ed. same year); also Rudolf Steiner College Publications. Sacramento, Calif. 1991]

[1959d]   *Darstellende Geometrie nach dynamischer Methode.* Verlag Freies Geistesleben. Stuttgart. [There is some reason to suspect that the book appeared initially in 1956. Large format. 55 pp. 167 figures, second edition 1982].

[1963]   *The International Waldorf School Movement.* Waldorf School Monographs. Englewood, N.J.

[1966]   *The Waldorf School Plan.* Waldorf School Monographs. Englewood, N.J. [also 1967, 3rd ed.]

[1967] *Introduction to Astronomy in the Sixth Grade of the Waldorf Schools*. Waldorf School Monographs. Englewood, N.J.

[1968] *Perspective Drawing*. Waldorf School Monographs. Englewood, N.J.

[1970] Conic Sections in Relation to Physics and Astronomy. *Math Teacher*. 63. pp. 101-109.

[1974] *Astronomy, an Introduction*. Waldorf School Monographs. [It seems that this was published by St. George Book Service, of Spring Valley, New York after acquiring the copyrights held by Waldorf School Monographs. As far as we know it was not part of the series of booklets published in Englewood, New Jersey. It is said that the text was found in a rough, unfinished state among Baravalle's papers. There was a second edition from Rudolf Steiner College, 1991, that reproduces [1974] without corrections.]

[1993] *On Teaching Mathematics and Physics*. Mercury Press. Spring Valley, New York. [English translation of *Zur Pädagogik der Physik und Mathematik*, 1921.]

[1994] *The Geometry of Shadow-Movements*. Mercury Press. Spring Valley, New York. [This English translation of [1952e] was bound together with D. J. van Bemelen, *A Drawing Lesson with Rudolf Steiner*, and is on pp.5–13 of a 13 page booklet.]

Other Publications

The following publications have been left out of the chronological list. Either the publications themselves were undated, or the editor has only seen them mentioned and no copies were available.

Durchblick durch die Erde. Verlag Freies Geistesleben. Stuttgart. [An edition is listed in the catalog of Emil Weises of Dresden in 1940. The first edition appeared before 1929 from Verlag der Naturwissenschaftlichen Sektion, Dornach, Switzerland.]

Die dynamische Schönheit geometrischer Formen.

Erweiterung der Perspektive Kümmerung des Sehraumes.
Waldorf-Verlag. Stuttgart. [We only know that it was before 1940 because it was listed for sale in 1940 by Verlag Emil Weises.]

Die Geometry des Pentagramms und der Goldene Schnitt. Stuttgart. [This title is listed by Verlag Emil Weises Buchhandlung in 1940. The publisher was Waldorf-Verlag, but there was no date printed on the booklet. The earliest dated edition that we have seen is that of J. Ch. Mellinger Verlag, 1950. Third edition 1969. Fourth ed. 1985. These differ significantly from the English article 1948b.]

Das Hervorgehen des Wissenschaftlichen aus dem Künstlerischen Einführung der Physik im 6 Schuljahr der Waldorfschulen. J. Ch. Mellinger-Verlag, Stuttgart.

Die Pädagogik Rudolf Steiners und die Erneuerung der deutschen Kultur. [This was also listed for sale by Verlag Emil Weiss Buchhandlungin 1940. We have not seen a copy in Spring Valley, however.]

Rudolf Steiner als Erzieher. J. Ch. Mellinger-Verlag, Stuttgart. [This may have appeared simultaneously with the English version [1952d].]

Index

Scientific, Mathematical, and Educational Books

available from

Parker Courtney Press
307 Hungry Hollow Road
Chestnut Ridge, New York
10977

Parker Courtney Press publishes for the Science and Mathematics
Association for Research and Teaching and prepares manuscripts for
other publishers.

Forming Concepts in Physics
by Georg Unger

Georg Unger's classic account of 20th century science has been
newly revised and translated for the first time into English. This
book gives an account of the astonishing revolutions in physics
during the first half of the twientieth century. The author gives
careful attention to the process of thought and imagination that
gives rise to physical concepts. The new role of probability theory
and its various interpretations are treated with unusual care. Unger
shares his thought process with the reader in a particularly
interesting and instructive way.
203 pages . . . . $ 24.75

Matter and Mind: Imaginative Participation in Science
by Stephen Edelglass, Georg Maier, Hans Gebert, and John Davy

This book develops an approach to science that does not look for
reality "behind" nature and recognizes the human knower within
nature.
"The short pithy text is likely to change many reader's
comprehension of science." – *Small Press Book Review*
"The authors of this small and remarkable volume work out of a
critique of the notion of objectivity....It is difficult to remember any
comparable scientific success." – R.H. Brady, *Holistic Education
Review.*
Lindisfarne, 136 pages $12.95

Finsler Set Theory: Platonism and Circularity
by David Booth and Renatus Ziegler

This book includes translations of Paul Finsler's most important papers on logic and set theory, a full historical account of the story of these suppressed classics of Platonism in mathematics, along with philosophical and mathematical essays by the authors. The recent revival of interest in non-well-founded sets is given thorough attention. Finsler, well-known as a differential geometer, was the heir to the Platonism of Cantor. He worked almost in isolation because of the domination of formalism during the years in which mathematical logic developed.

Birkhäuser, 268 pages .   .   .   . (available in 1996)

To Order: Send a check to Parker Courtney Press at the address on the previous page. Include $3.50 for shipping. Orders to New York State require sales tax.